PROJEKTRAUM

Roman Kurzmeyer

Erlebte Modelle
Model Experience

Renate Buser
Marie Sester
Elizabeth Wright
Katharina Grosse
Susanne Fankhauser
Sarah Rossiter
Lee Bul
Martina Klein
@ home
Heinz Brand
Eran Schaerf
Anya Gallaccio

Projektraum
Kunsthalle Bern 1998–2000

Edition Voldemeer Zürich

SpringerWienNewYork

Roman Kurzmeyer, Basel

Die vorliegende Publikation dokumentiert die Ausstellungen
im Projektraum der Kunsthalle Bern 1998–2000.

Förderung des Gesamtprojekts:
Alfred Richterich Stiftung, Laufen
Förderung der Publikation:
Alfred Richterich Stiftung, Laufen
Fondation Nestlé pour l'art, Lausanne
Förderung einzelner Veranstaltungen:
Schweizer Kulturstiftung Pro Helvetia / Galerie Tschudi, Glarus / Fondation Nestlé
pour l'art, Lausanne / The British Council, Switzerland / Ambassade de France, Bern /
Teo Jakob AG, Bern

Veröffentlicht in einer Auflage von 1200 Exemplaren, davon 48 Exemplare als Vorzugsausgabe, numeriert von 1 bis 48. Die Exemplare der Vorzugsausgabe wurden von Renate Buser,
Marie Sester, Elizabeth Wright, Katharina Grosse, Susanne Fankhauser, Sarah Rossiter,
Lee Bul, Martina Klein, @home, Heinz Brand, Eran Schaerf und Anya Gallaccio bearbeitet,
ergänzt und signiert.

*Published in an edition of 1200 copies, including 48 copies presented as a special edition
and numbered 1 through 48. The books comprising the special edition have been personally
reviewed and signed by Renate Buser, Marie Sester, Elizabeth Wright, Katharina Grosse,
Susanne Fankhauser, Sarah Rossiter, Lee Bul, Martina Klein, @home, Heinz Brand, Eran
Schaerf and Anya Gallaccio and contain additions and modifications made by the artists.*

Dieses Exemplar trägt die Nummer
This copy bears the number

Photograph der Ausstellungen: Michael Fontana, Basel / Übersetzung: John S. Southard,
Groß-Umstadt / Umschlagfoto: Huang Qi, Zürich / Lektorat: Julia Hagenberg, Köln /
Gestaltung: Voldemeer AG, Zürich / Satz: Marco Morgenthaler, Zürich / Repro: tnt graphics,
Bassersdorf / Druck und Bindung: Stämpfli AG, Bern
Gedruckt auf säurefreiem, chlorfrei gebleichtem Papier – TCF

 Voldemeer AG
Postfach
CH-8039 Zürich

Printed in Switzerland

SPIN 10764274

Mit 78 Abbildungen

ISBN 3-211-83469-9 Springer-Verlag Wien New York

SpringerWienNewYork, Sachsenplatz 4–6, A-1201 Wien

Inhalt / Contents

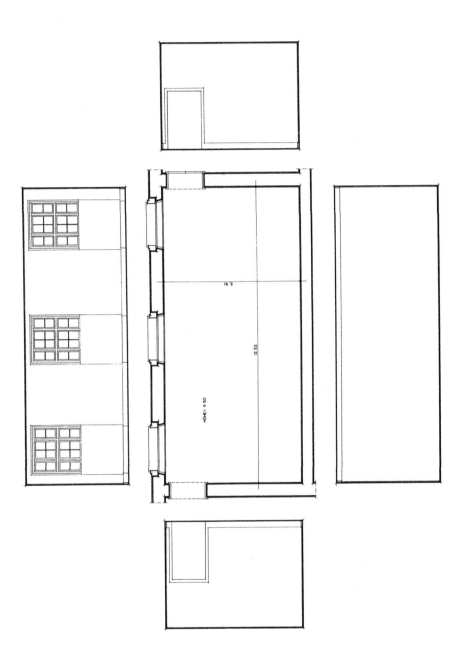

Als Projektraum diente der Seitenlichtsaal Ost im Tiefparterre der Kunsthalle Bern.
The "Projektraum" occupied the rear hall of the basement floor of the Kunsthalle Bern.

Vorwort

Eine Kunsthalle hat die Aufgabe, aktuelle Entwicklungen und Debatten in der zeitgenössischen Kunst kritisch zu reflektieren und aktiv mitzugestalten. Retrospektiv ausgerichtet und zu einer kulturellen Gesamtschau verpflichtet sind die Kunstmuseen. Von einer Kunsthalle wird erwartet, daß sie neue Werke zur Diskussion stellt und als Partnerin der Kunstschaffenden die Rolle der Produzentin übernimmt. Im Rahmen der finanziellen und räumlichen Möglichkeiten haben wir während der vergangenen zwei Jahre im Projektraum versucht, neben Ausstellungen, die künstlerische Positionen zeigten, die in der Schweiz noch nicht oder kaum bekannt waren, auch Ausstellungen zu veranstalten, in denen von uns angeregte Arbeiten zu sehen waren. Unser Interesse galt aktuellen künstlerischen Positionen und weniger bestimmten Medien, Diskursen oder künstlerischen Bewegungen. Das vorliegende Buch dokumentiert und kommentiert die Ausstellungen einzeln und in chronologischer Reihenfolge. Den Abbildungsteilen sind Texte beigeordnet, die die Ausstellungen lesbar zu machen versuchen. Verfaßt wurden die einzelnen, monographisch ausgerichteten Kapitel, während die entsprechende Ausstellung zu sehen war. Alle Beiträge wurden später für dieses Buch überarbeitet und erweitert. Die Dokumentation der Ausstellungen verdanken wir dem Photographen Michael Fontana, der mit Umsicht und Verständnis im Projektraum gearbeitet hat.

=

Mein Dank gilt Bernhard Fibicher, Direktor der Kunsthalle Bern, auf dessen Einladung ich ab 1998 für zwei Jahre den Projektraum leitete. Danken möchte ich auch Werner Schmied, unter dessen sorgfältiger und überlegter technischer Leitung die Ausstellungen aufgebaut wurden, sowie allen weiteren Mitarbeiterinnen und Mitarbeitern der Kunsthalle Bern. Die Schweizer Kulturstiftung Pro Helvetia ermöglichte den Vortrag von Sun Jung Kim, Leiterin des Artsonje Center Seoul, die anläßlich der Präsentation dieses Buches über die Zukunft der Kunst im Zeitalter der Globalisierung sprach. Unsere Tätigkeit wurde permanent und substantiell durch die Alfred Richterich Stiftung gefördert. Ich bedanke mich bei Alfred Richterich herzlich für sein Vertrauen. Mein besonderer Dank gilt allen Künstlerinnen und Künstlern, die sich für dieses Projekt einsetzten.

R. K.

Foreword

The task of the public art gallery, known in Switzerland as the "Kunsthalle", is to pursue critical enquiry into current developments and discussions in contemporary art and to take an active part in shaping them. Whereas art museums are dedicated to providing a retrospective view and illuminating a broader cultural context, a public art gallery is expected to present new works for discussion and to act as a producer and a partner to artists. To the extent possible within the framework of our budgetary and spatial resources, we at the "Projektraum" (Project Space, the German name of the project program and the exhibition space at the Kunsthalle Bern) have attempted to present not only exhibitions intended to highlight positions of artists previously unknown or relatively little known in Switzerland but also shows featuring works created in response to our own suggestions. Our interest has focused more closely on current artistic positions than on specific media, discourses or contemporary movements in art. This book documents and offers commentary on the exhibitions presented during the past two years, individually and in chronological order. The illustrations are accompanied by texts which should help readers to develop a better understanding of the exhibitions. Each chapter is a monograph devoted to a specific exhibition and written while the exhibition was in progress. All of the articles were edited and expanded for this book. Exhibition documentation was provided by the photographer Michael Fontana, who brought a great deal of thoughtful insight to his work at the "Projektraum".

=

I wish to thank Bernhard Fibicher, Director of the Kunsthalle Bern, who invited me to serve as director of the "Projektraum" for two years beginning in 1998. I would also like to thank Werner Schmied, who provided invaluable guidance for the realisation of the exhibitions as Technical Director, and the entire staff of the Kunsthalle Bern. The Arts Council of Switzerland Pro Helvetia sponsored the lecture by Sun Jung Kim, Director of the Artsonje Center Seoul, who spoke about the future of art in the era of globalization at the presentation ceremony for this book. We received continuous and substantial support in our work from the Alfred Richterich Stiftung. I am very grateful to Alfred Richterich for his trust. Above all, I wish to thank all of the artists for their commitment to this project.

R. K.

Erlebte Modelle

Als der amerikanische Künstler Richard Serra 1969 in der Kunsthalle Bern ein *Splashing Piece* realisierte, schickte er sich an, die Ortsspezifik der Skulptur neu zu definieren. Die Ausstellung, in der Serras Arbeit zu sehen war, trug den Titel *When Attitudes Become Form* und zählt heute zu den wichtigsten Ausstellungen der westlichen Nachkriegskunst. Bern diskutierte die Ausstellung als Ärgernis. Harald Szeemann, seit 1961 für die Programmgestaltung der Kunsthalle Bern verantwortlich, sah das Vertrauen in ihn und seine Kunstauffassung erschüttert und verließ das Haus. Wie schon im Jahr zuvor in einem von der Galerie Castelli in Manhattan benutzten Lagerhaus hatte Richard Serra auch in Bern 210 kg geschmolzenes Blei in der Eingangshalle in den Winkel zwischen Boden und Wandansatz geworfen und erhärten lassen. Die Arbeit bestand aus der Handlung, die das Werk an jener Stelle ermöglichte, und dem Ort, von dem es nicht entfernt werden konnte, ohne zerstört zu werden. Richard Serra stand in dieser Ausstellung mit seiner Werkauffassung nicht allein da, der deutsche Plastiker Joseph Beuys etwa schuf in der Kunsthalle Bern mit seiner Fettecke ebenfalls ein ortsspezifisches Werk. Die radikalste Geste aber war sicher diejenige des Amerikaners Michael Heizer, der den Gehsteig vor der Kunsthalle mit einer Abbruchkugel bearbeitete. »Wirkliche Ortsspezifik«, schreibt der Kunstwissenschaftler Douglas Crimp, »ist immer eine politische Spezifik.« Aus seiner Perspektive wären die heftigen Reaktionen, die diese Ausstellung insgesamt und insbesondere Arbeiten wie *Berne Depression* von Michael Heizer auslösten, Beleg für die Öffentlichkeit, die es den Künstlern herzustellen gelang, und damit für die Qualität ihrer Arbeiten. Crimp sieht als Konsequenz dieser Ortsspezifik, daß der Betrachter zum Subjekt des Werkes wird. Der Ort des Werkes erwies sich nun als ein Raum, den der Betrachter begehen konnte. In der Wahrnehmung dieses Ortes ist der Künstler dem Betrachter nicht überlegen: Das Werk ist der Ort, der Ort ist das Werk. Ein Kritikpunkt der damaligen Werkauffassung besteht nach Crimp allerdings darin, daß der Ort nur im formalen Sinne als spezifisch verstanden wurde. Jeder Ort konnte deshalb ein einzigartiger Ort sein, selbst ein Galerie-Raum, ein Werk deshalb auch an unterschiedlichen Orten wieder ausgeführt werden.

=

Eine Bedingung für die von der amerikanischen Kunst der sechziger Jahre vertretene Ortsspezifik ist das gegenstandsfreie Bild und somit eine Bildform, die in der westlichen Kunst dem 20. Jahrhundert angehört. Der Bildort war seit der Erfindung der Zentralperspektive in der Malerei der Renaissance bis ins späte 19. Jahrhundert ein dargestellter Ort. Vorausgesetzt wurde, daß jedes Werk einen Referenten hat. Das Kunstwerk suggerierte, daß der Bildort Abbild eines außerhalb des Werkes existierenden Ortes sei; daran erinnert selbst noch das knappe Stück Boden bei Rodins *Bürgern von Calais* (1889). Mit der Gegenstandslosigkeit war eine Problematik entstanden, die – unabhängig davon, wie deren Bedeutung heute für das Selbstverständnis unserer Epoche eingeschätzt wird – von den Künstlern selbst gesehen und behandelt, in der Rezeption ihrer Werke aber oft vernachlässigt wurde. Eine Konsequenz dieser Sensibilisierung für Fragen des Ortes war die Installationskunst der sechziger und siebziger Jahre, deren beste Arbeiten für den Ort ihrer Präsentation geschaffen wurden. Die Künstler begannen später außerhalb der Museen Ausstellungen zu organisieren, im Außenraum der Städte, in der Landschaft, in verlassenen Industriebauten, Bahnhöfen, Kirchen und schließlich auch an privaten Orten wie Wohnungen oder Kellern. Diese Entwicklung hatte Auswirkungen auf die Arbeit in den traditionellen Institutionen, in denen man nun unter einer Ausstellung die Inszenierung von Kunstwerken an einem bestimmten Ort oder unter spezifischen räumlichen Bedingungen zu verstehen begann.

=

Werke wie Richard Serras *Splashing Piece* wurden an unterschiedlichen Orten und unter sehr verschiedenen Rahmenbedingungen ausgeführt. Man hat in der Handhabung dieser Arbeiten als autonome Werke die Preisgabe von deren eigentlicher Intention vermutet, doch problematisch ist nicht, daß Künstler verschiedene Fassungen ihrer ortsspezifischen Werke realisieren, sondern daß in der Diskussion dieser Werke von deren Struktur geredet wird, als existiere die einzelne Arbeit ohne den Ort der Ausführung. Selbstverständlich ist das im Betrachter erzeugte Bild nicht dasselbe, wenn er das erwähnte Werk Serras 1968 in Manhattan in einem Lagerhaus oder 1969 in Bern in der Kunsthalle gesehen hat. Zu unterscheiden ist nicht nur zwischen dem Bildort und dem Ort des Werkes, die bei der eben diskutierten Arbeit von Richard Serra zusammenfallen, sondern auch dem Ort des Bildes, das ein Werk im Betrachter erzeugt. Mit der Bezeichnung dieses dritten Ortes ist das Verhältnis von Bild und Werk angesprochen, neben produktionsästhetischen Überlegungen werden in der weiteren Argumentation deshalb auch stärker rezeptionsästhetische Gedanken hinzugezogen.

=

Wenn ich nun auf die Gegenwart zu sprechen komme, so im Bewußtsein dieser Klammer, die meine bisherigen Ausführungen geöffnet haben. Seit Richard Serra 1969 in der Kunsthalle Bern heißes Blei warf, haben viele Künstler und einige Künstlerinnen in diesen Räumen ihre Werke ausgestellt. Jeder Leiter gab der Kunst seiner eigenen Zeit Raum und favorisierte dabei sehr unterschiedliche Traditionen innerhalb der Kunst des 20. Jahrhunderts. Eine Kunsthalle ist keiner bestimmten Werkauffassung oder künstlerischen Haltung verpflichtet, sondern sie ist eine Institution, welche die Veränderungen in der zeitgenössischen Kunst kritisch begleitet. Es ist bemerkenswert, daß der Debatte um die Ortsspezifik während der letzten dreißig Jahren immer besonderes Gewicht zukam, obschon die Programme der verschiedenen Leiter sehr unterschiedlich waren. In modifizierter Form beschäftigte auch mich die Frage der Ortsspezifik bei der Planung der in diesem Buch vorgestellten Ausstellungsreihe: Meine Aufmerksamkeit galt dem Ort, den Kunstwerke als Bild unserer Vorstellung oder als Erfahrung erzeugen. Richard Serras damalige Auffassung von Ortsspezifik vermochte sich nicht durchzusetzen, nicht einmal der Künstler selbst hielt an ihr fest. Die Erfahrung aber, daß erst die Wechselbeziehung von Objekt, Kontext und Betrachter eine Arbeit als Kunstwerk konstituiert, ist seither aus der Praxis der Kunst nicht mehr wegzudenken.

=

Auf die Frage nach meinen Auswahlkriterien antwortete ich einmal, der Ausgangspunkt meiner Beschäftigung mit einem künstlerischen Werk sei immer die Verbindung eines Originals mit meiner eigenen Realität. Ich hätte auch sagen können, Kunst sei die Realität jener Menschen, die von ihr angesprochen und berührt werden und sich mit ihr beschäftigen. Von Kunst sprechen, dies wollte ich mit meiner Antwort ausdrücken, heißt jene Wechselbeziehung zur Sprache bringen, die sich zwischen einer künstlerischen Arbeit und dem Betrachter vollziehen kann, wobei die jeweilige Verständnisbereitschaft und -möglichkeit für die Entfaltung eines Werks eine bestimmende Rolle spielen. Die Kunst liegt denn auch nicht in der Form, und doch hat diese für die Wirkung von Malerei, Skulptur, Zeichnung, Video oder Installation einen genauso hohen Stellenwert wie beispielsweise für gute Architektur. Daß das Kunstwerk über die Form allein nicht erfaßt werden kann, liegt daran, daß dessen Form nicht Zweck, sondern Ergebnis von Grundlagenarbeit in einem bestimmten, vom Künstler bearbeiteten Untersuchungsfeld ist. Auf was ich anspiele, hat der amerikanische Anthropologe Gregory Bateson einst auf die einfache Formel gebracht, die Karte sei nicht das Territorium. Die Karte gibt Informationen, die Orientierung und sichere Bewegung im Raum ermöglichen soll. Sie beschreibt den Raum mittels der Angabe von Unterschieden. Ein Blick in die Vergangenheit zeigt, wie sich die Karte, also die Auffassung von Raum, verändert hat, auch dann, wenn das Territorium selbst das gleiche geblieben ist. Dies liegt nicht nur daran, daß sich die Kartographie weiterentwickelt hat, sondern auch an den im Verlauf der Zeit wechselnden Auffassungen darüber, welche Informationen so bedeutsam seien, daß sie für den Kartenleser einen Unterschied machten. Künstler arbeiten auf der Ebene der Karte, im Unterschied etwa zum Architekten, der baubare und benutzbare Lösungen für konkrete räumliche Situationen entwickeln muß. Kunstwerke sind Modelle von Wirklichkeit. Kunstwerke sind Form und verlangen nach der Erfahrung ihrer Wahrnehmung. Ein Bild allerdings entsteht erst dann, wenn der Betrachter wahrnimmt, wie Kunst auf die Welt trifft.

=

Ein Kunstwerk braucht keinen Referenten, um als solches erkannt zu werden. Zwar behandelt die Bildtheorie auch die Bildsprache, doch bezweifle ich, daß das Wesen der Kunst in ihrer Sprachlichkeit begründet liegt. Kunst ist nicht in erster Linie ein Kommunikationsmittel und unterscheidet sich darin von allen anderen visuellen Medien. Wer Kunst als Kommunikationsmittel auffaßt, der wird sich vor allem dann von einem Kunstwerk angesprochen fühlen, wenn dieses sich auf eine bekannte Formensprache, einen Stil oder einen Diskurs bezieht. Die Stärke des Kunstwerks liegt in dem, was der französische Kunstwissenschaftler Georges Didi-Huberman als »intensive Form« bezeichnet hat. In Anlehnung an den deutschen Kunstkritiker Carl Einstein schreibt er, der Ursprung der Präsenz einer Form sei »im Spiel ihrer Formation und ihrer Präsentation und nicht nur in ihrem Symbolismus« zu suchen. Er beschreibt das Kunstwerk als etwas Sichtbares, das sich in die »tiefe Einsamkeit seiner Form« zurückziehe. Intensive Formen sprechen uns in ihrem Für-sich-Sein und durch ihre Singularität an. Es sind Konfigurationen, die sich unserer Wahrnehmung als Eigenformen offenbaren. Die intensive Form ist kein Bild, sondern ein Ort ohne einfache Verweisstruktur.

=

Diese Überlegungen rufen mir einen Ausstellungsbesuch von 1995 in Washington D.C. in Erinnerung. In der Sammlung der National Gallery of Art, wo ich Leonardos um 1478 gemaltes Bildnis der Ginevra de' Benci aufsuchte, war gleichzeitig eine kleine Sonderausstellung zu sehen, die Modelle und Zeichnungen zu drei bedeutenden Bauten der Renaissance, dem Dom von Pavia, der Peterskirche zu Rom und dem Dom Santa Maria del Fiore in Florenz, versammelte. Daß in Italien seit dem 15. Jahrhundert das Zeichnen von Plänen und das Bauen maßstabsgerechter, geschnitzter und bemalter Modelle zu den üblichen Planungsarbeiten der Architekten zählten, wissen wir aus den Schriften von Leon Battista Alberti. Für die Künstler, die diese Modelle nach ihren Vorstellungen und Zeichnungen anfertigen ließen, waren die Modelle Instrumente, um ein genaueres Bild dessen zu bekommen, was sie in ihren Plänen geistig angelegt hatten, aber auch, um die Gunst der Bauherren zu gewinnen und später den Handwerkern für deren Arbeit eine Orientierungshilfe zu geben. Interessanterweise aber gehen viele dieser Modelle in Anlage und künstlerischer Ausgestaltung weiter als die schließlich realisierten Gebäude. Sie sind Zeugnisse architektonischer Visionen und ungebauter, manchmal unbaubarer Architektur. Die Modelle haben heute den Status autonomer künstlerischer Werke. Daß es sich vielleicht schon für die Zeitgenossen so verhalten hat, läßt das Beispiel des Modells für Antonio da Sangallos Peterskirche in Rom vermuten, an dem der Assistent Antonio Labacco sieben Jahre arbeitete und dessen Kosten die Aufwendungen für den Bau einer Kirche überstiegen.

=

Ein Modell ist nicht nur Vorbild oder Nachbildung, die plastische Darstellung eines Werkes in verkleinertem Maßstab, sondern auch dessen Urform. Jedes Modell steht in diesem doppelten Bezugssystem, womit auch klar wird, weshalb dieser Begriff für die Beschäftigung mit Werken der Kunst so wertvoll ist. Das Kunstwerk ist Nachbildung, insoweit es geschaffene Formen, Stoffe und Ordnungen wiederholt, und Vorbild, insofern es erlebt werden will. Seine Wirkung entfaltet sich erst im Betrachten, Lesen oder Zuhören. Das Kunstwerk ist widersprüchlich: Es gleicht einer Übertragung, zu der die Vorlage erst noch zu ermitteln ist. Meine Gedanken zu den im Projektraum ausgestellten Arbeiten sind ein Versuch in diese Richtung. Von der Idee des ortsspezifischen Werkes der sechziger Jahre führte dieser Beitrag zur Beschreibung des Ortes als einer zusammengesetzten Einheit, die sich durch die Wechselbeziehung von Objekt, Kontext und Betrachter bildet. Die in diesem Band besprochenen Ausstellungen machen anschaulich, inwiefern dieser Ort in der gegenwärtigen künstlerischen Praxis eine wichtige Bezugsgröße ist.

Model Experience

With his Splashing Piece, *a work realised at the Kunsthalle Bern in
1969, the American artist Richard Serra redefined the site-specific
character of sculpture. The exhibition at which Serra's work was shown
bore the title "When Attitudes Become Form" and is regarded today as one
of the most important exhibitions of post-war Western art. The exhibition
was seen in Bern as an affront. Harald Szeemann, Director of the Kunsthalle Bern since 1961, took the negative reaction as a vote of "no con-
fidence" in himself and his approach to art and left the institution. In an
action performed the preceding year at a show presented in a warehouse
used by the Castelli Gallery in Manhattan, Richard Serra cast 210 kg of
molten lead into the corner formed by the floor and the wall in the foyer
and let it harden. Serra's work consisted of the act itself, which made it
possible to realise the work at that specific site, from which it could not be
removed without being destroyed. Richard Serra was not alone with his
work concept at this exhibition. The German sculptor Joseph Beuys also
created a site-specific work at the Kunsthalle Bern – his* Fettecke (Lard
Corner). *Surely the most radical gesture of all, however, was that of the
American artist Michael Heizer, who attacked the sidewalk in front of the
Kunsthalle with a demolition ball. "Arbitrary site-specificity", writes the
art scholar Douglas Crimp, "is always political specificity". Viewed from
his perspective, the heated reactions triggered by this exhibition as a whole
and especially by such works as Michael Heizer's* Berne Depression *were
an affirmation of the publicity the artists succeeded in attracting and thus
of the quality of their works as well. According to Crimp, one of the con-
sequences of this site-specificity is that the viewer becomes a subject of the
work. The site of the work was now identified as a space which the viewer
could enter. In terms of the experience of this place, the artist has no ad-
vantage over the viewer. The work is the place; the place is the work. One
of the problems Crimp sees in connection with the concept of the work expressed at the time, however, is that place or site was understood as specific only in formal terms. Thus every place could be unique, even a room
in a gallery, which also means that a work could be realised at different
places.*

=

*One of the preconditions of the site-specificity advocated by American
artists of the 1960s is the non-representational pictorial image – that
is, a pictorial form that is specific to 20th-century Western art. From the invention of central perspective in Renaissance painting to the late 19th century, pictorial place was a depicted place. This presupposed the existence
of a referent for every work of art. The work implied that the pictorial place*

was the image of a place that existed outside the work itself. We are re-
minded of this by such things as the small patch of ground in Rodin's Citi-
zens of Calais *(1889). Non-representational art posed a set of problems*
which – regardless of current assessments of its significance to the self-
image of our era – was recognised and addressed by artists themselves
but often overlooked in the critical reception of their works. One outgrowth
of this increased concern with matters of place was the installation art of
the 1960s and 1970s, the best examples of which were created specifically
for the site at which they were presented. Later on, artists began organis-
ing exhibitions in places other than museums, in urban outdoor space, in
the countryside, in abandoned industrial buildings, railway stations and
ultimately at private locations, such as flats or cellars. This had an impact
on the work of traditional institutions as well, where the idea of an exhibi-
tion as the staged presentation of works of art at a specific site or under
specific spatial circumstances found increasing acceptance.

=

Works like Richard Serra's Splashing Piece *were realised at many*
different sites and in diverse environments. Some assumed that treat-
ing these works as autonomous works of art would compromise their actual
intent, but the problem is not that artists realise different versions of their
site-specific works but that discussion of these works focuses on their
structure, as if the individual work of art could exist without the site at
which it is realised. It goes without saying that the image created in the
mind of the viewer will be different depending upon whether he saw Serra's
work in a Manhattan warehouse in 1968 or in Bern at the Kunsthalle in
1969. It is important to distinguish not only between the pictorial place and
the site of the work, which coincide in the case of the work by Serra dis-
cussed above, but between pictorial place and the image generated by a
work in the mind of the viewer. With the identification of this third notion
of place, we address the relationship between the visual image and the
work. In addition to considerations pertinent to the aesthetics of produc-
tion, the following argumentation will therefore focus more specifically
upon the aesthetics of reception as well.

=

In returning to present-day matters, I have not forgotten the thought
with which the preceding remarks began. In the years since Richard
Serra cast lead on the floor at the Kunsthalle Bern, many male artists and
several women artists have exhibited their works here. All of the directors
made space available for the art of their respective times, each giving pref-
erence to very different traditions of 20th*-century art. A public art gallery*
has no obligations of loyalty to any particular artistic attitude or approach
to works of art. Its duty as an institution is to observe and follow changes
in contemporary art with a critical eye. In view of the fact that the programs

developed by the various directors differed considerably, the special emphasis placed on the debate regarding site-specificity over the past thirty years seems quite remarkable. In somewhat modified form, the issue of site-specificity was of concern to me as well in planning the series of exhibitions presented in this book. The focus of my attention was the sense of place evoked by the work of art as an image in our minds or as experience. Richard Serra's concept of site-specificity has not endured, and even Serra himself eventually abandoned it. However, the recognition that a work becomes a work of art only through the interplay of object, context and viewer has become indispensable to the practice of art today.

=

Asked about my selection criteria, I once replied that the point of departure for my interest in a work of art is always the link between an original and my own reality. I could just as easily have said that art is reality for those it touches and speaks to, for people who concern themselves with it. What I wanted to put across with my answer was the idea that to talk about art is to address the reciprocal relationship that can develop between a work of art and the viewer, whereby the willingness and ability of a given viewer to comprehend plays a very definite role. Art is not found in form, although form is as important to the impact of painting, sculpture, drawing, video art or installation as it is to that of good architecture. The reason why a work of art cannot be grasped through form alone is that its form is not an end in itself but rather the result of basic research in a field of investigation chosen by the artist. The point I am trying to make has been reduced to a simple formula by the American anthropologist Gregory Bateson: the map is not the territory. A map gives information that is intended to provide orientation and reliable keys for safe movement within a space. It describes the space by pointing out differences. A look into the past shows that the map – the view of space – has changed, even where the territory itself remains the same. This is true not only because cartography has progressed but also because opinions on the question of what information is so important that it makes a difference to the map reader have also changed over the course of time. Artists work at the level of the map, unlike architects, for instance, who must first develop constructable and viable solutions for concrete spatial situations. Works of art are models of reality. Works of art are form, and they demand the experience of perception. A picture does not emerge, however, until the viewer experiences art in its encounter with the world.

=

A work of art needs no referent in order to be recognised as such. Although pictorial theory also deals with pictorial language, I seriously doubt that the essence of art lies in its linguistic character. Art is not primarily a means of communication, and that distinguishes it from other *visual media. Those who see art as a vehicle for communication are most likely to respond to a work of art that can be related to a familiar style, discourse or formal language. The strength of the work of art is to be found in what the French art scholar Georges Didi-Hubermann refers to as "intensive form". With reference to the German art critic Carl Einstein, he writes that the origin of the presence must be sought "in the interplay of its formation and its presentation and not in its symbolism". Didi-Hubermann describes the work of art as something visible which retreats "into the profound solitude of its form". Intensive forms speak to us in their self-containedness and through their singularity. They are configurations which reveal themselves to our perception as autonomous forms. The intensive form is not an image or a picture but a place that has no simple referential structure.*

=

These thoughts remind me of a visit I made to an exhibition in Washington, D.C. in 1995. The National Gallery of Art, where I viewed Leonardo's painted portrait of Ginevra de Benci, a work painted in 1478, was also showing a small special exhibition featuring models and drawings of three major Renaissance buildings – the cathedral of Pavia, the Church of St. Peter in Rome and the Santa Maria del Fiore Cathedral in Florence. We know from the writings of Leon Battista Alberti that architects routinely drafted plans and built carved, painted scale models for planning purposes. For the artists who executed these models on the basis of their ideas and drawings, the models were tools which enabled them to form a more exact picture of the intellectual concepts defined in their plans, but they also helped them gain favour with their clients and provide guidelines for the craftsmen who carried out the work. It is interesting to note, however, that many of these works are more sophisticated – in terms of both conception and artistic design – than the actual buildings themselves. They are documents of architectural visions and unbuilt, often unbuildable architecture. Such models are regarded as autonomous works of art today. That contemporaries may well have viewed them in the same way is suggested, for example, by the model for Antonio de Sangallo's Church of St. Peter in Rome, a piece on which Sangallo's assistant Antonio Labacco worked for seven years at a cost that exceeded the construction budget for a church.

=

A model is more than an original source or a replica, the three-dimensional representation of a work on a reduced scale. It is the primal form of the work of art. Every model has its place in this dual reference system, a fact which also explains why the term is so valuable to those who study works of art. The work is a replica to the extent that it repeats forms, contents and ordering systems and a source insofar as it demands to be experienced. Its effects unfold only through observation, reading or listening. The work of art involves a contradiction. It is like a translation for which the original remains unknown. My thoughts about the works exhibited in the "Projektraum" represent an attempt to pursue this line of enquiry. This essay proceeds from the idea of the site-specific work of the 1960s to a description of place as a complex whole formed by the inter-relationships that link the object, the context and the viewer. The exhibitions discussed in this volume are clearly indicative of the importance this concept of place has achieved as a reference value in the contemporary practice of art.

Echo

Renate Buser
Marie Sester
Elizabeth Wright

Elizabeth Wright nähte 1991 ein Abendkleid und einen Pelzmantel für einen Bücherstoß. Seither kopiert sie in leicht verändertem Maßstab weitverbreitete, unauffällige Dinge, die ihr dennoch irgendwo aufgefallen sind und sich ihr eingeprägt haben. Sie verkleinert oder vergrößert Möbel, Kleidungsstücke, Bücher, Gebäude, Akten, Autos und Fahrräder sowie seit kurzem auch Spuren, die bremsende Fahrzeuge auf der Fahrbahn zurückgelassen haben. Ließen ihre frühen Arbeiten eine surrealistische Weltsicht vermuten, so zeigt sich nun ein photographischer Blick in der Erfassung von Wirklichkeit: Das Kunstwerk dient der Beglaubigung ihrer visuellen Erinnerung. Die in Handarbeit nach photographischen Vorlagen hergestellten Objekte sind im eigentlichen Sinne Artefakte, ihre visuelle Präsenz aber soll sich um nichts von derjenigen ihrer Vorbilder unterscheiden. Anders als die Künstler in der Duchamp-Nachfolge zeigt Elizabeth Wright diese Nachbildungen nicht als Kunstwerke, sondern sucht für diese einen Ort, an dem die Werke die Funktion jenes Gegenstandes übernehmen, dem sie nachempfunden sind. Im vorliegenden Falle war es ein verbogenes, auf dem Gehsteig vor der Kunsthalle liegendes Fahrradrad, angekettet an der Abschrankung zum Helvetiaplatz. *Stolen Bicycle* (1998), ausgestellt als Kunstwerk im Kunstkontext, wäre belanglos. Wirkung können diese Art von Arbeiten nur entfalten, wenn sie in den Außenraum versetzt werden. Dort, am Originalschauplatz, übernehmen sie die Funktion eines Motivs und werden vom Publikum entweder übersehen oder besonders beachtet. Wer der Erscheinungsform des vergrößerten Rades vertraute, für den wurde die Vergrößerung zum Maßstab der Realität. Das Rad bestimmte einen Ort, um den herum sich eine Spielzeugwelt aufbaute und bewegte. Seit den frühen neunziger Jahren stehen Fragen des Kontextes im Mittelpunkt des Interesses von vielen jungen Künstlerinnen und Künstlern. Die Arbeiten von Elizabeth Wright zeigen, daß auch die Dekontextualisierung ein aufschlußreiches Verfahren ist, um zu einem an-gemessenen Verständnis von Wirklichkeit zu gelangen.

=

Diese These sollte im Innern der Kunsthalle Widerhall finden und durch die Kombination einer Raumarbeit von Marie Sester mit Fotoarbeiten von Renate Buser bekräftigt werden. Renate Buser reiste 1996 auf dem Frachtschiff *MS Sachsen* in zwölf Tagen von Livorno nach Montreal.

Auf dieser Reise entstanden Photographien, die das Leben an Bord, die See und den Horizont dokumentieren – ein Tagebuch. Die Bilder und Erzählungen beziehen sich auf ihre Erfahrungen von unterwegs. Kürzlich hat mir Renate Buser einen Auszug aus einem Vortrag des französischen Philosophen und Historikers Michel Foucault aus dem Jahre 1967 zugeschickt, in dessen Worten sie jene Erfahrungen gespiegelt sieht: »Bordelle und Kolonien sind zwei extreme Typen der Heterotopie, und wenn man daran denkt, daß das Schiff ein schaukelndes Stück Raum ist, ein Ort ohne Ort, der aus sich selber lebt, der in sich geschlossen ist und gleichzeitig dem Unendlichen des Meeres ausgeliefert ist und der, von Hafen zu Hafen, von Ladung zu Ladung, von Bordell zu Bordell, bis zu den Kolonien suchen fährt, was sie an Kostbarstem in ihren Gärten bergen, dann versteht man, warum das Schiff für unsere Zivilisation vom 16. Jahrhundert bis in unsere Tage nicht nur das größte Instrument der wirtschaftlichen Entwicklung gewesen ist (nicht davon spreche ich heute), sondern auch das größte Imaginationsarsenal. Das Schiff, das ist die Heterotopie schlechthin. In den Zivilisationen ohne Schiff versiegen die Träume, die Spionage ersetzt das Abenteuer und die Polizei die Freibeuter.« In Montreal ging Renate Buser an Land. Sie kaufte einen Wagen, um die Stadt zu erkunden. Wieder war sie unterwegs, diesmal auf den Straßen einer Großstadt. Nach der langen Reise, während der das Auge in der endlos scheinenden Weite ohne Orientierung geblieben war, rückte ihr die Welt nun buchstäblich auf den Leib. Renate Buser photographierte in Montreal als Fremde. Sie erschloß sich die Stadt von außen. Sie zeigt deren Ansicht. Der damals entstandenen Serie von Hochhausfassaden hat sie den Titel *Objects in the mirror may be closer than they appear* (1997) gegeben. Der Titel bezieht sich auf einen Schriftzug, eine Warnung, auf dem Rückspiegel des Fahrzeuges, mit dem sie unterwegs war. Die Bilder von Renate Buser basieren auf digital unbearbeiteten Photographien und entstehen im Labor. Die Künstlerin sieht das Bild nicht im Sucher ihrer Kamera, sondern schafft es bei der Arbeit mit den Negativen. Von den konzeptuellen Vorgaben abgesehen, ist es vor allem dieser Prozeß des Vergrößerns, in dessen Verlauf die meisten künstlerischen Entscheide fallen. Es resultieren großformatige Handvergrößerungen auf Barytpapier. Jedes Bild ist ein Unikat, obschon es sich um eine Photographie handelt. Auf den Bildern aus dem Stadtraum Montreal, die Renate Buser in ihren Ausstellungen wie Plakate direkt auf die Wand klebt, erscheinen die Bauten aus ihrem städtebaulichen Umfeld herausgelöst und in fiktiven Konstellationen. Die Fassadenausschnitte sind als Flächen aufgefaßt und als solche zueinander in Beziehung gesetzt. Die Gebäude wirken entrückt und schwerelos. Zugleich erschließt Buser durch die Dekontextualisierung der Bauten eine andere Ebene ihrer Realität, die dem Architekturphotographen verschlossen bliebe: Sichtbar wird das Verhältnis zwischen vorgegebener architektonischer Struktur und individueller

Nutzung. Der Blick auf die Fassaden ist zugleich auch ein Blick hinter die Fassaden.

=

Der Beitrag von Marie Sester zur Ausstellung in Bern war eine Plexiglasstruktur, die die seit dem Bauhaus standardisierten Dimensionen einer Einzimmerwohnung im Maßstab 1:1 nachbildet, modellhaft nachvollziehbar macht und dadurch der Kritik unterwirft. Die Installation *Appartement* (1996) versetzte den Betrachter in drei verschiedene Positionen: Die Enge dieser Wohnung war von innen als Raumabfolge körperlich erlebbar und optisch von außen als Volumen sowie von oben als gezeichnete Struktur wahrnehmbar. Durchdrang der Blick die gläsernen Wände, traf er nicht nur auf die realen Mauern des Ausstellungsraumes, der das Werk enthielt, sondern auch auf die wie eine zweite Haut auf diese Wände aufgezogenen Photographien, was, wo auch immer man sich im Saal aufhielt, den Eindruck verstärkte, zugleich innen und außen zu sein. Marie Sester hat Architektur studiert, sich aber schon während des Studiums entschlossen, nie ein Haus zu bauen. Als Künstlerin beschäftigt sie sich dennoch immer mit Fragen, die das Verhältnis von westlicher Architektur, Geschichte und Gesellschaft betreffen. Eine ihrer seither entstandenen Arbeiten etwa handelt von der globalen, grenzenlosen Gesellschaft, die sie als Mythos entlarvt. Wiederum findet sie in der Architektur die Belege für ihre These. Die Grenzen, so die Künstlerin, wurden nicht aufgehoben, sondern lediglich ins Innere der Flughäfen verlegt, zu denen nur Zutritt erhält, wer über Paß, Visum und Geld verfügt. Zwischen der Grenz- und Sicherheitskontrolle vor dem Abflug in unserer eigenen Gegenwart und der Kontrolle an einem Stadttor in einer früheren Epoche besteht für Marie Sester kein grundsätzlicher Unterschied.

=

Die Kombination der Werke von Buser und Sester in einem fensterlosen Raum setzte einen schwindelerregenden Wechselwirkungsprozeß in Gang, der es noch schwieriger machte, einen eigenen Standort gegenüber den Hochhäusern zu bestimmen. Der fiktive, durch die plakatierten Fassaden erzeugte Stadtraum erwies sich als ein Raum ohne Ort. Halt und Orientierung wurden zu einer Sache des Innenraums, der persönlichen Geschichten, die sich hinter den Fassaden abspielen, sichtbar für alle, die nahe vor die Bilder traten. Diese Beobachtung läßt mich noch einmal zurückkommen auf den Vortrag von Michel Foucault, den mir Renate Buser auszugsweise zugeschickt hat. 1982 sagte Michel Foucault über den in seinem Vortrag zentralen Begriff der Heterotopie, er verstehe darunter »jene Art von Räumen, denen inmitten eines sozial definierten Bereichs eine ganz andere Funktion eignet, die sogar in völligem Widerspruch zu denen des umliegenden Raums stehen« könne. Michel Foucault hatte seinen Vortrag von 1967 als Text in seinem Todesjahr 1984 autorisiert; doch in der Kunst wirksam geworden sind seine Gedanken – wie Daniel Defert am Beispiel des kubanischen Künstlers Felix Gonzalez-Torres ausführte, der 1991 in Manhattan auf großformatigen Reklameflächen die nicht für den öffentlichen Gebrauch bestimmte Photographie eines ungemachten Bettes zeigte – erst in unserer Gegenwart.

Echo

Renate Buser
Marie Sester
Elizabeth Wright

In 1991, Elizabeth Wright made an evening dress and a fur coat for a stack of books. Since then, she has made copies, to slightly modified scale, of a number of commonly available, unremarkable items that have nevertheless caught her attention at one place or another and made an impression on her. She reduces or enlarges furniture, items of clothing, books, buildings, files, automobiles, bicycles and, more recently, tyre marks left behind on roadways by braking cars. Whereas her earlier works suggest a Surrealist view of the world, we now recognise the workings of a photographic eye in her handling of reality. The work of art serves as an affirmation of her visual experience. The hand-made objects produced from photographs are artefacts in the true sense of the word, although their visual presence is not meant to differ in any way from that of her models. Unlike artists still working in the tradition of Duchamp, Elizabeth Wright does not present these replicas as works of art but instead selects a place for them in which they can assume the functions of the original objects they are based upon. The work presented in Bern was an enlarged, bent-up bicycle wheel lying on the sidewalk in front of the Kunsthalle and chained to the barrier along Helvetiaplatz. Exhibited as a work of art in the context of art, Stolen Bicycle (1998) would have been totally inconsequential. Works of this kind achieve striking effects only when they are placed outdoors. There, in their original setting, they take on the function of a motif and are either overlooked by the public or become objects of special attention. For those who accepted the formal manifestation of the bicycle wheel, the enlargement became the measure of reality. The wheel took command of a site around which a toy world emerged and began to revolve. Since the early 90s, many young artists have focused attention on issues of context. The works of Elizabeth Wright show that the technique of decontextualisation can also be a revealing means of attaining an appropriate understanding of reality.

=

This idea was to be echoed in the interior of the Kunsthalle and emphasised through the combination of a spatial installation by Marie Sester and photo pieces by Renate Buser. In 1996, Renate Buser took a twelve-day voyage from Livorno to Montreal on the freighter "MS Sachsen". In the course of her journey, she took photographs that document life

Elizabeth Wright
Stolen Bicycle, 1998

Elizabeth Wright
Stolen Bicycle, 1998

Renate Buser
Westmount Square, Montreal, 1997

Marie Sester
Appartement, 1996

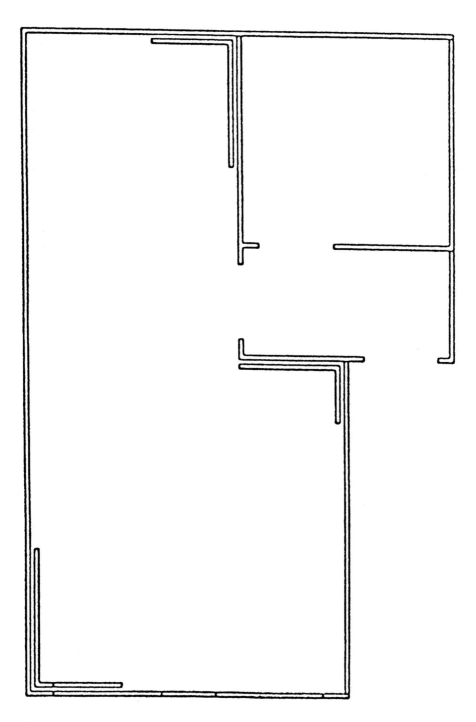

Marie Sester
Appartement, 1996, Grundriß

Renate Buser
Crescent Hotel, La Canadienne, Westmount Square, Montreal, 1997

30

Marie Sester
Appartement, 1996

Marie Sester
Appartement, 1996

Renate Buser
Westmount Square, Montreal, 1997

Renate Buser
Marriott Hotel, Montreal, 1997

Marie Sester
Appartement, 1996

34

Katharina Grosse
Inversion, 1998

36

Katharina Grosse
Inversion, 1998

Katharina Grosse
Inversion, 1998

on board ship, the ocean and the horizon – a diary of sorts. The artist's photos and comments relate to her experiences during the crossing. Renate Buser recently sent me an excerpt from a lecture presented in 1967 by the French philosopher and historian Michel Foucault, in whose remarks she saw her own experience reflected: "Brothels and colonies are two extreme types of heterotopism, and if one realises that the ship is a rocking chunk of space, a place without location that lives from itself, self-enclosed yet at the same time abandoned to the infinitude of the sea, and travels from port to port, from cargo to cargo, from brothel to brothel and on to the colonies in search of the most precious treasures hidden in their gardens, then one begins to understand why the ship has not only been the greatest instrument of economic development for our civilisation from the 16th century to our day (which is not my topic here) but also the greatest arsenal of the imagination. Indeed, the ship is the epitome of heterotopism. Dreams faded and disappeared in the civilisations that had no ships; adventure gave way to espionage, privateers to the police." *Renate Buser landed in Montreal, where she bought a car and began to tour the city. She was underway once again, this time on the streets of a metropolis. After her long sea voyage, during which the eye found no orientation in the apparently limitless expanse, she was now literally embraced by the world. Renate Buser photographed as a stranger in Montreal. She approached the city from the outside. What she shows is its face. The title she gave to the series of high-rise facades completed at the time was* Objects in the mirror may be closer than they appear *(1997), a phrase borrowed from the warning printed on the rear-view mirror in her rented car. Renate Buser's pictures are based upon digitally unprocessed photographs and took shape in the darkroom. She does not see the image in the camera viewfinder; she creates it in her work with the negatives. Underlying conceptual principles notwithstanding, it is during this enlargement process that most of her artistic decisions are made. The products are large-scale, hand-made enlargements on baryta paper. Although a photograph, each picture is a one-of-a-kind piece. In the photos of downtown Montreal, which Renate Buser applies directly to the walls of exhibition rooms like posters, the buildings appear to have been removed from their urban surroundings and placed in fictitious constellations. The details of facades are seen as flat surfaces and set as such in relationship to one another. The buildings have a displaced, weightless look. By decontextualising these architectural structures, Buser also presents them at a different level of reality that is not accessible to architectural photographers. The relationship between existing architectural structure and individual use becomes apparent. The view of the facades is also a view behind the facades.*

=

Marie Sester's contribution to the exhibition in Bern was a Plexiglas structure as a replica of a one-room flat in the dimensions established as standard by the Bauhaus on a scale of 1:1, which made it possible to reconstruct these dimensions in the form of a model, thus exposing them to critical scrutiny. The installation Appartement *(1996) placed the viewer in three different positions. The confining space of the flat could be experienced physically from the inside as a spatial sequence, visually from the outside as a volume and from above as a drawn structure. Viewers gazing through the glass walls saw not only the real walls of the exhibition room in which the work was installed but also caught a glimpse of the photographs that adhered to those walls like a second skin, an effect which heightened the impression of being inside and outside at the same time. Marie Sester studied architecture but decided before graduating that she would never build a house. As an artist, however, she is still concerned with the interrelationships of Western architecture, history and society. One of her later works deals with the global, borderless society, which she exposes as a myth. Here again, the artist finds evidence for her thesis in architecture. The borders, she contends, have not been eliminated but simply shifted to places inside airports to which only those with passports, visas and money are permitted entry. Marie Sester sees no essential difference between present-day pre-flight border and security checks and the control measures carried out at the gates of cities in bygone eras.*

=

The combination of works by Buser and Sester in a windowless room generates a baffling process of interaction which makes it even more difficult for viewers to establish their position vis-à-vis the high-rise buildings. The fictional city space defined by the postered facades proved to be a place without location. Stability and orientation became a matter of inner space, of the personal stories that unfolded behind the facades, visible to all who took a closer look at the pictures. This observation brings me once again to the lecture by Michel Foucault to which Renate Buser originally called my attention. Speaking in 1982 on the meaning of heterotopism, the central concept of his earlier lecture, Foucault said that he understood it to denote "the kind of spaces that, situated within a socially defined area, take on entirely different functions that may even stand in total opposition to those of their surroundings". Although Michel Foucault authorised the written version of his 1967 lecture in 1984, the year of his death, the influence of his thoughts on art did not become evident until our decade, as Daniel Defert has pointed out with reference to the Cuban artist Felix Gonzalez-Torres, who presented the photograph of an unmade bed, an image not intended for public consumption, on billboards in Manhattan in 1991.

Der Malerei gewordene Blick

Katharina Grosse

Im Verlauf des 20. Jahrhunderts wurde keine künstlerische Disziplin so oft totgesagt wie die Malerei. Einzelnen Künstlerinnen und Künstlern ist es dennoch gelungen, mit ihrer Malerei die Debatte neu zu beleben und eigenständige Positionen zu bestimmen. In den späten siebziger Jahren drehte sich die Diskussion für kurze Zeit um Formen expressiver und figurativer Malerei, die allerdings sofort in Verdacht standen, hinter den komplexen Bildbegriff der klassischen Avantgarde zurückzufallen. Seit den achtziger Jahren gilt die Aufmerksamkeit der Kritik vor allem der analytischen Malerei, die nach den essentiellen Eigenschaften der Malerei als Medium fragt und diese als Malerei thematisiert. Vorbehalte beziehen sich hier meistens auf die mit der Metadiskursivität einhergehenden Ausblendung der visuellen Repräsentation, aber auch der Autorschaft. Die Kriterien der analytischen Malerei – die malerische Befragung von Farbe, Träger, Auftrag – sind zwar auch Merkmale der Malerei von Katharina Grosse, allerdings benutzt die Künstlerin lediglich deren Ergebnisse, nicht aber die Methode. Katharina Grosse malt auf Leinwand, Papier, Aluminium und direkt auf die Wand. Ihre Arbeiten sind klar und einfach strukturiert; innerhalb dieser funktionalen Bildanlage versucht sie, ein möglichst großes Maß an malerischer Freiheit zu erlangen. Die Herausforderung liegt in der Übersetzung von Energie in Malerei. Die Malbewegung mit breiten Pinseln oder der Sprühpistole spielt dabei eine zentrale Rolle. Die Malerei von Katharina Grosse läßt eine Disposition für schnelles, situationsgerechtes und dem inneren Antrieb verpflichtetes malerisches Handeln erkennen. Ihre Malerei lebt vom Widerspruch zwischen streng formaler Bildanlage und betont freiem Zugriff auf diese Bildanlage. Die Reinheit der Mittel und der Verzicht auf Darstellung haben lediglich instrumentelle Bedeutung und meinen weder Beschränkung noch Ausschließung, sondern Konzentration auf die grundlegenden Mittel der Malerei.

=

Katharina Grosse kam nach Bern, um ein Wandbild zu sprühen. Absicht der Malerin war es, eine Situation zu schaffen, in der ein Ort der Malerei in relativer Unabhängigkeit zum gegebenen architektonischen Raum wahrnehmbar würde. Die Malerei sollte verweisen, ohne abzubilden, zu imitieren oder darzustellen. Das in der Kunsthalle Bern nach langen Gesprächen schnell und konzentriert gesprühte Wandbild *Inversion* (1998), ein Werk auf Zeit und als solches geplant, benutzte zwar die Architektur als Träger, war aber nicht als architekturbezogene Arbeit angelegt. Es handelt sich um Katharina Grosses erste Arbeit, die die Künstlerin ausschließlich mit der Sprühpistole ausführte. Sie sprühte grüne Acrylfarbe in eine Raumecke ohne Vorzeichnung direkt auf Wand und Decke. Die hohen Bäume und dichten Büsche vor den drei großen Fenstern des Projektraumes ließen schon das in den Seitenlichtsaal einfallende Licht grün erscheinen. Die von der Künstlerin verwendete Farbe konnotierte zwar Natur, doch der Blick aus den Fenstern auf das Grün der Bäume und Büsche im Außenraum rückte die von Katharina Grosse ausgewählte Farbe zurück in den Kunstbereich. Sichtbar waren die durch das alte Mauerwerk vorgegebenen Unebenheiten der Wandflächen, die Struktur der Tapete und die Risse im Mauerwerk. Die Malerei war faktisch. Sie bildete die Wand in ihrer Oberfläche ab und war zugleich bestimmte Form an einem bestimmten Ort.

=

Sprüharbeiten haben per se ein unarchitektonisches Raumverhalten, da unter großem Druck zerstäubte Farbe nur bedingt kontrolliert werden kann und feinste Farbpartikel sich überall im Raum absetzen. Die gesprühte Farbe verbindet sich nicht mit dem gestrichenen Untergrund, sondern sitzt tropfenförmig auf der Wand. Aus großer Distanz wirkte die Farbe der Arbeit von Katharina Grosses dennoch, als ob sie aus der Wand dränge. Der Farbauftrag war von unterschiedlicher Dichte, nie aber völlig deckend. Wahrnehmbar blieb, wie die Malerin die Sprühpistole geführt hatte, nämlich auf der Stelle kreisend oder horizontal bewegt. Tropfen bildeten kleine Wolken. Die Wolken überlagerten sich zu einem Fleck, dessen Umriß an einigen Stellen eher linear, an anderen unscharf, weich und flächig wirkte. Die Arbeit erinnerte an Leonardos Aufforderung zum Träumen mit offenen Augen: »Und das geschieht«, heißt es in dessen Schriften, »wenn du manches Gemäuer mit verschiedenen Flecken oder mit einem Gemisch aus verschiedenartigen Steinen anschaust; wenn du dir gerade eine Landschaft ausdenken sollst, so kannst du dort Bilder verschiedener Landschaften mit Bergen, Flüssen, Felsen, Bäumen, großen Ebenen, Tälern und Hügeln verschiedener Arten sehen; ebenso kannst du dort verschiedene Schlachten und Gestalten mit lebhaften Gebärden, seltsame Gesichter und Gewänder und unendlich viele Dinge sehen, die du dann in vollendeter Form und guter Gestalt wiedergeben kannst.« Der flaschengrüne Fleck regte in vergleichbarer Art zum Sehen an, war zugleich aber auch die beabsichtigte Form. Die Wand wurde als Oberfläche an einer bestimmten Stelle sichtbar gemacht. Stirnwand, Längswand und Decke treffen dort zusammen. Wer sich nahe vor die Malerei begab, wurde von ihr umfangen. Dunkel und drohend saß sie über dem Kopf des Betrachters auf den Mauern, beim ersten Anblick dagegen wirkte die Arbeit klein und erschien weit weg. Die Arbeit befand sich schräg gegenüber dem Eingang. Leere umgab sie. Wer den Saal betrat, war mit dem Werk unmittelbar konfrontiert. Die Malerei besetzte jene Stelle im Raum, auf die das Auge als erste fiel: Die Malerei verursachte ein Bild, vergleichbar einem Bild gewordenen Blick auf die Wand.

=

Katharina Grosse sprühte in der Kunsthalle Bern zum ersten Mal eine Wandarbeit; wenig später sollte sie im Kunstverein Bremerhaven erstmals bemalte Papiere hinzuziehen, die sie ausdrücklich für andere räumliche Zusammenhänge konzipiert hatte. An den ausgesparten Stellen im Papier war erkennbar, daß die Arbeiten für andere architektonische Situationen entstanden waren. Katharina Grosse zeigte raumbezogene als autonome Arbeiten und löste damit das Problem der visuellen Repräsentation, ohne die reine Flächigkeit ihrer Arbeit – seit Cézanne das Signum von Modernität – aufzugeben, ohne darzustellen, ohne zu imitieren oder abzubilden. Deutlicher noch als frühere Ausstellungen thematisierte jene in Bremerhaven den Unterschied zwischen dem Ort der Malerei und dem Ort des Bildes. Der Ort der Malerei ist das Papier oder die Wand: die Fläche, nicht der Raum. Die Malerei bildet keinen illusionären Ort. Indem Grosse für ihre Ausstellung in Bremerhaven Arbeiten heranzog, die raumspezifisch entstanden waren, verschärfte sie den in ihren Sprüharbeiten angelegten Widerspruch zwischen raumbezogener und autonomer Arbeit. Die kontroversen Reaktionen, die ihre erste gesprühte Wandarbeit in der Kunsthalle Bern ausgelöst hat, lassen sich nun besser erklären. Katharina Grosse hatte mit ihrer Arbeit das ungeschriebene Gesetz verletzt, wonach raumspezifische Interventionen den Ort als architektonischen Raum reflektieren und als solchen zeigen müssen. Bestimmend für die Konzeption dieser Arbeit war aber nicht der erfahrene, erlebte und analysierte Raum, sondern die Perspektive des Betrachters, der den Raum betrat. Der Ort der Malerei fällt zusammen mit dem Ort, auf den der Blick des eintretenden Betrachters trifft. Der Ort des Bildes dagegen war einer der Imagination, erzeugt durch das Verhältnis von individueller Betrachterdisposition, Werkbegrenzung, Farbe und Architektur.

The View Transformed into Painting

Katharina Grosse

In the course of the 20th century, no other artistic discipline has been declared dead as often as painting. Nevertheless, a number of artists have succeeded in reviving the debate and in defining independent positions with their painting. During the 1970s, discussion focused briefly on forms of expressive and figurative painting, although these were soon suspected of lagging far behind the complex pictorial concept of the classical avant-garde. Since the 1980s, critical attention has been devoted above all to analytical painting, to its concern with the essential characteristics of painting as a medium and to the attempt to deal with such themes in painting. Reservations expressed in this context usually relate to the exclusion of visual representation as a consequence of the focus on metadiscourse but also refer to issues of authorship as well. While the criteria of analytical painting – the analysis of colour, image-bearing media and application – clearly characterise the painting of Katharina Grosse, this artist uses only their results, not the method itself. Katharina Grosse paints on canvas, paper, aluminium and directly on walls. Her works are clearly and simply structured. Within this functional pictorial scheme she seeks to attain the greatest possible measure of painterly freedom. The challenge lies in the translation of energy into painting, and her painting technique, using broad brushes or a spraygun, plays a crucial role in this context. Clearly evident in Katharina Grosse's painting is a disposition toward rapid, situation-oriented painting actions that respond to an inner driving force. Her painting derives its power from the tension between a strict formal pictorial scheme and an emphatically free approach to dealing with it. In her work, purity of artistic means and absence of representational intent are meaningful only in an instrumental sense; signifying neither restriction nor exclusion, they are indicative of complete concentration on the fundamental resources of painting.

=

Katharina Grosse came to Bern to do a mural with spray paint. Her intention was to create a situation in which a painting could be experienced in relative independence from existing architectural space. The painting was to allude without depicting, imitating or representing. Executed rapidly and with great concentration at the Kunsthalle Bern after a series of lengthy discussions, the spray-painted mural entitled Inversion *(1998), a temporary work that had been planned as such, made use of the architecture as a medium but was not conceived as a piece relating to architecture. It was the first painting ever done by Katharina Grosse using only a spraygun. She sprayed green acrylic paint directly onto the wall and ceiling in the corner of a room, without a preliminary drawing. The tall trees and dense shrubbery outside the three large windows of the "Projekt-raum" gave a green cast to the light entering the room known as the "Sei-tenlichtsaal". Although the colour the artist selected suggested nature, the view of the green trees and bushes outside the windows qualified this impression, placing the artist's chosen colour within the realm of art. The irregularities in the surfaces of the walls covering the old masonry, the structure of the wallpaper and the cracks in the masonry itself remained visible. The painting was factual. It presented an image of the surface of the walls yet established its identity as a specific form at a specific place.*

=

Spray-painted works exhibit a non-architectural spatial behaviour per se, as paint sprayed under high pressure cannot be controlled completely, and because fine particles tend to distribute themselves throughout the surrounding space. The sprayed paint does not bond with the painted groundcoat but attaches itself in drops to the walls. Viewed from a distance, however, the paint used in Katharina Grosse's work appeared to seep from inside the wall. The paint was applied in varying density but did not form an opaque cover anywhere. It was possible to retrace the painter's motions with the spraygun – circular movements from a standing position and horizontal swaths applied while moving along the wall. Drops formed small clouds. The clouds overlapped, creating a spot, the outline of which appeared to form a line at some points, becoming diffuse, soft and expansive at others. The work called to mind Leonardo's idea of dreaming with open eyes: "And that occurs", he wrote, "when you look at walls that have different spots in them or a combination of different kinds of stones: If it is a landscape you wish to imagine, then you can see pictures of different landscapes there, with mountains, rivers, rocks, trees, broad plains, valley and hills of many different varieties; and you see diverse battles and figures with animated gestures, strange faces and garments and an infinite number of things which you can then recreate in complete form and fine appearance." The bottle-green spot stimulated vision in a similar way, yet it was also the specific form intended. The wall was made visible as surface at a specific place, the point at which front wall, side wall and ceiling meet. Those who came close to the painting were surrounded by it. Dark and threatening, it hung on the walls above the viewer's head, despite the fact that, at first glance, the work looked small and appeared to be far off in the distance. The painting was positioned diagonally opposite to the entrance. It was surrounded by emptiness. Those who entered the room were immediately confronted by it, as the painting occupied the point in the room to which the eye was first drawn. The painting generated an image, much like a view of a wall that has become a painting.

Katharina Grosse did her first spray-painted wall piece at the Kunsthalle Bern. Somewhat later, she presented paintings on paper at the Kunstverein Bremerhaven, works she had expressly conceived for quite different spatial contexts. The cut-out spaces in the paper indicated that the works had been done for other architectural settings. Katharina Grosse exhibited works with a specific spatial orientation as autonomous pieces, thus solving the problem of visual representation without sacrificing the pure flat-surface quality of her work – the emblem of modernity since Cézanne – and without representation, imitation or depiction. In an even more striking manner than earlier exhibitions, the Bremerhaven presentation focused on the difference between the place represented in the painting and place as it takes shape in the image in the viewer's mind. The place of the painting is the paper or the wall – the flat surface and not the space. The painting creates no illusionary space. In presenting works conceived for a specific spatial setting at her exhibition in Bremerhaven, Grosse underscored the opposition between site-oriented and autonomous art that underlay her spray-painted works. The controversy sparked by her first spray-painted wall piece at the Kunsthalle Bern is now somewhat easier to explain. With her work, Katharina Grosse had violated the unwritten law that site-specific interventions must reflect and show place as architectural space. The determining factor in the conception of this work was not perceived, experienced and analysed space but rather the perspective of the viewer entering the space. The place of the painting coincided with the place to which the gaze of the viewer entering the space was drawn. The place in the image in viewer's mind, on the other hand, was an imaginary one, a product of the relationships between the individual disposition of the viewer, the boundaries of the work, colour and architecture.

Das Museum der Tiere

Susanne Fankhauser

Die Zeichnung *Das Museum der Tiere* von Susanne Fankhauser ist ein monumentaler Ink-Jet-Druck von 300 x 965 cm auf Kunststoff. Die für unsere Ausstellung konzipierte Arbeit ist klein, auf Bildschirmgröße erarbeitet und für die Präsentation im Projektraum industriell auf die Maße der Längswand gegenüber der Fensterfront vergrößert worden. Das Arbeitsmaterial der Künstlerin waren wie schon bei ihrer Rauminstallation *Die Betrachter und das Kunstwerk* (1996) Abbildungen von zeitgenössischen Kunstwerken. Wiederum hat sie ausschließlich Arbeiten verwendet, die schon reproduziert vorlagen. Die Künstlerin findet ihre Motive in Zeitungen, Zeitschriften, Katalogen und Büchern: Sie bearbeitet Ausschnitte der Reproduktionen am Computer. »Durch die stark stilisierende bildnerische Bearbeitung«, schreibt die Künstlerin Christa Ziegler zur Arbeitsweise von Susanne Fankhauser, »werden die Elemente zu gleichgestellten Spielsteinen.« Obschon Susanne Fankhauser in ihrer neuen Arbeit Abbildungen plastischer Werke verwendet, gilt ihre gestalterische Aufmerksamkeit besonders den Konturen und den Flächen ihrer Figuren.

=

Die Künstlerin sagt von der in Bern ausgestellten Zeichnung, sie ermögliche wie ein großes Fenster den Blick in ein Museum. Das Museum, in das Fankhauser den Betrachter führt, ist allerdings kein Gebäude. Es sind auch keine Menschen zu sehen: Kunstwerk und Betrachter lassen sich in diesem Bild-Museum nicht voneinander unterscheiden. Die Kunstwerke selbst bilden und sind das Museum. Sowohl die ausgestellten Kunstwerke als auch die betrachtenden Tiere sind Abbildungen künstlerischer Werke: Susanne Fankhauser zeigt in ihrer Zeichnung unter anderem Arbeiten von Braco Dimitrijevic, Mike Kelley, Lothar Baumgarten, Claudia Di Gallo, Jochen Gerz, Maurizio Cattelan, Katharina Fritsch, Robert Rauschenberg, Marie José Burki, Jeff Koons, Wim Delvoye, Charles Ray, Stephan Balkenhol, Paul Thek, Bruce Nauman, Abigail Lane, Ashley Bickerton, Joseph Beuys und Diego Giacometti. Den Titel *Das Museum der Tiere* hat sie um die mit den Reproduktionen aufgefundenen Angaben zu Künstler, Werktitel und Entstehungsjahr ergänzt. Die Sammlung, die Susanne Fankhauser zeigt, setzt sich aus Arbeiten von Künstlerinnen und Künstlern unterschiedlicher Bedeutung zusammen. Aufnahme in die Sammlung haben nicht etwa Kunstwerke gefunden, die von der Kunstkritik, der Kunstgeschichtsschreibung oder der Künstlerin persönlich besonders geschätzt werden, maßgebend für die Auswahl war vielmehr die künstlerische Zielsetzung, in ihrer Zeichnung möglichst interessante Konfigurationen zu schaffen. Da Fankhauser mit Abbildungen arbeitet, war der durch den jeweiligen Photographen festgehaltene Blick auf ein Werk für die Auswahl entscheidend. Hätte die Künstlerin eine Arbeit in einer anderen Abbildung kennengelernt, hätte sie diese womöglich nicht in ihre Zeichnung aufgenommen, bestimmt aber nicht in jene Nachbarschaft kopiert, in der wir sie nun sehen. Susanne Fankhauser ist wie wir alle eine Kunstbetrachterin. Als Künstlerin ist sie zusätzlich eine Betrachterin der Kunstbetrachtung. Als Künstlerin, die mit Kunstwerken arbeitet, interessiert sie sich für deren konkrete Erscheinung und nur am Rande für die künstlerischen Konzepte, aus denen die Arbeiten hervorgegangen sind. Verzerrungen und Entstellungen, wie sie durch die photographische Reproduktion entstehen, sieht sie nicht als Fehler oder Differenz zur Vorlage, sondern im Gegenteil als das, was diese tatsächlich auch sind: Formen. Die Reproduktion weist bei Fankhauser nicht zurück auf ein Original. Die Reproduktion ist das Original, das die Künstlerin als Form interessiert.

=

Susanne Fankhauser suggeriert in ihrer Zeichnung eine Abfolge von Räumen. Sie tut dies allein mittels der Anordnung der Figuren. Sie verzichtet in ihrer Zeichnung auf Böden, Wände und Decken. Da, wo sie Wände eingezogen hat, waren diese schon auf der verarbeiteten Reproduktion Bestandteil des abgebildeten Werkes. Auf den ersten Blick scheint alles seine Ordnung zu haben, doch sobald das Auge durch das Bild zu wandern und die fiktive räumliche Situation zu erkunden beginnt, erweist sich diese Welt als endlos, bodenlos und orientierungslos. Die Elemente, die den Bildraum konstituieren, sind abzählbar. Die Farben auf dieser Zeichnung sind bunt, grell, unmoduliert. Sie halten Distanz zur Natur. Obwohl die abgebildeten Artefakte nicht nach den Regeln perspektivischer Darstellung angeordnet wurden, erzeugen die Beziehungen zwischen den Bildelementen Konfigurationen, die vom Auge als auszulotender Raum wahrgenommen werden. Die Anordnung und die wechselseitigen Beziehungen zwischen den abgebildeten Werken sind wichtiger als die einzelnen Bildfiguren. Das Einzelne ist auf dieser Zeichnung nur in der Kombination mit allen anderen Elementen von Bedeutung.

=

Der philosophische Relativismus besagt, daß nur die Verhältnisse der Dinge zueinander, nicht aber diese selbst erkennbar seien. Die Zeichnung von Susanne Fankhauser erinnert in dieser Hinsicht an *Das imaginäre Museum* des Spanienkämpfers und späteren französischen Kulturministers André Malraux (1901–1976). Dieses Museum ohne Mauern umfaßt alle Kunstwerke, die, wie der Kunstwissenschaftler Douglas Crimp schreibt, »der mechanischen Reproduktion unterworfen werden können und somit der diskursiven Praxis, welche die mechanische Reproduktion möglich gemacht hat: die Kunstgeschichte«. Kunst, wie wir sie heute in unserer Kultur verstehen, entstand erst im 19. Jahrhundert, gleichzeitig mit der Photographie, dem Museum und der Kunstgeschichte als universitärem Fach. Das Museum machte aus einem höfischen Bildnis ein Gemälde und aus einem kirchlichen Ritualgegenstand eine Skulptur: Es löst die Werke aus allen räumlichen, sozialen und historischen Bindungen und ihren angestammten Funktionen, um sie zu ästhetischen Gegenständen zu erklären. André Malraux erkannte, daß das Verfahren der photographischen Reproduktion ermöglichen würde, erstmals überhaupt in der Menschheitsgeschichte eine Sammlung aufzubauen, die allen Menschen zugänglich sein würde und in der sich Werke verschiedenster Epochen und Kulturen vergleichen ließen. Die Vereinigung aller Meisterwerke in einem Museum war sein Wunsch, sein Projekt erinnert denn auch an die Idee des Gesamtkunstwerks. Malraux war sich bewußt, daß die Reproduktion das Werk nicht nur aus seinem Kontext herauslöst, sondern zusätzlich auch seiner Stofflichkeit beraubt. Zwar zeigt er sich davon überzeugt, daß sein Museum ohne Mauern die Intellektualisierung zum Äußersten treibe, doch lassen seine Formulierungen keinen Zweifel daran, daß jedes Stück seiner Sammlung einen Referenten in der Welt der Dinge hat. Das Museum ohne Mauern von Malraux ist ein Inventar und eine Phänomenologie. Seit André Malraux sein Museum entworfen und durch seine Publikationen für das Publikum zugänglich gemacht hat, ist die Reproduktionstechnik durch die neuen Medien revolutioniert worden. Der technische Fortschritt in der Bildverarbeitung ermöglicht Bilder, die wie Reproduktionen wirken, aber Erfindungen sind. Der Verwendungsmodus der Reproduktionen im Schaffen von Susanne Fankhauser zeigt, daß die Künstlerin Original und Reproduktion unterschiedlichen Ordnungen zuzählt. Ihr Museum zählt zur Welt der Bilder. Der Referent eines bearbeiteten Bildes ist in jedem Fall die Reproduktion und nicht deren Vorlage. Die Reproduktion wird nicht verwendet, um eine Wirklichkeit außerhalb der Darstellung zu repräsentieren, sondern um einen imaginären Ort zu schaffen.

The Animal Museum

Susanne Fankhauser

Susanne Fankhauser's drawing entitled Das Museum der Tiere *(The Animal Museum) is a monumental inkjet print on plastic measuring 300 x 965 cm. The work conceived for our exhibition is small. It was executed in computer-screen size and industrially enlarged to match the dimensions of the long wall opposite the front windows for presentation at the Kunsthalle. As in her spatial installation* Die Betrachter und das Kunstwerk *(The Viewer and the Work of Art, 1996) the artist used photographs of contemporary artworks in this piece, incorporating once again only works of which reproductions were already available. She finds her motifs in newspapers, magazines, catalogues and books and processes details from the reproductions on the computer. "By virtue of her the highly stylised approach to image-processing", writes the artist Christa Ziegler with reference to Susanne Fankhauser's working method, "the elements become playing pieces of equal value". Although Susanne Fankhauser makes use of photos of sculptural works in her new piece, her creative interest is focused primarily on the contours and surfaces of her figures.*
=
With regard to the drawing exhibited in Bern, the artist says that, much like a large window, it opens a view into the museum. The museum Fankhauser invites her viewers to enter is not a building, however, and there are no people there. It is impossible to distinguish between the viewer and the work of art in this museum of images. The artworks themselves form the museum – indeed, they are *the museum. Both the works of art presented and the animals shown looking at them are images reproduced from existing works of art. In her drawing, Susanne Fankhauser includes works from such artists as Braco Dimitrijevic, Mike Kelley, Lothar Baumgarten, Claudia Di Gallo, Jochen Gerz, Maurizio Cattelan, Katharina Fritsch, Robert Rauschenberg, Marie José Burki, Jeff Koons, Wim Delvoye, Charles Ray, Stephan Balkenhol, Paul Thek, Bruce Nauman, Abigail Lane, Ashley Bickerton, Joseph Beuys and Diego Giacometti. To her title,* Das Museum der Tiere, *she added data on the artist, work title and year of completion provided with each of the reproductions (some of which may be incorrect). The collection Susanne Fankhauser presents is composed of works by artists of differing significance. It was not restricted to artworks singled out by art critics, art historians or herself as particularly worthy of attention. Instead, she selected works primarily on the basis of their suitability to her objective of creating the most interesting configurations possible in her drawing. Because Fankhauser works with photos, the view of a*

given work presented by the respective photographer was crucial to the selection process. Had she found a particular work in a different photograph, she might well have left it out of her drawing and would certainly have copied it into a different neighbourhood of images than the one in which we now see it. Like everyone else, Susanne Fankhauser is a viewer of art. As an artist, she is also an observer of the process of viewing art. As an artist who works with works of art, she is interested in their concrete manifestation and only marginally in the artistic concepts from which they have emerged. She does not regard the distortions and changes that result from the process of photographic reproduction as faults or deviations from the original but as the very thing that they indeed are: forms. In Fankhauser's art, the reproduction does not refer back to the original; it is an original in its own right and of interest to the artist as form.

=

Susanne Fankhauser suggests a sequence of spaces in her drawing, relying solely on the arrangement of he figures to accomplish this. She adds no floors, walls or ceilings to her drawing. The walls she does include were already present in the work as represented in the photographs she processed. At first glance, everything appears governed by a system of order, but as soon as the eye begins to wander through the picture, exploring the fictional spatial situation, the world it discovers proves to be endless, bottomless and devoid of orientation. The elements that constitute the pictorial space are countable. The colours in the drawing are varied, vivid and unmodulated. They keep nature at a distance. Although the artefacts represented are not arranged in accordance with the rules of perspective, the relationships among the pictorial elements nevertheless produce configurations which the eye perceives as space for exploration. The arrangement and the interrelationships among the photographed works are more important than the individual pictorial figures. In this drawing, the individual is of significance only in combination with all of the other elements.

=

Philosophical relativism tells us that only the relationships among things are evident, and not the things themselves. Viewed from this vantage point, Susanne Fankhauser's drawing calls to mind The Imaginary Museum *of the erstwhile Spanish-Civil-War soldier and later French Minister of Culture André Malraux (1901–1976). His museum without walls encompasses all works of art which, as art scholar Douglas Crimp has noted, "can be subjected to the process of mechanical reproduction and thus to the discourse that mechanical reproduction made possible: art history". Art as it is now understood in our culture did not emerge until the 19th century with the advent of photography, the museum and art history as a course of university studies. The museum transformed the courtly portrait into a painting, the object of religious ritual into a sculpture. It* released objects from all existing spatial, social and historical bonds and stripped them of their established functions in order to declare them works of art. André Malraux recognised that the process of photographic reproduction would make it possible, for the first time in the history of mankind, to build a collection accessible to all in which works from a variety of epochs and cultures could be compared with one another. His great desire was to have all masterworks united in a single museum, and his project reminds us of the idea of the gesamtkunstwerk – the total work of art. Malraux knew that reproduction not only took the work from its context but also robbed it of its material properties as well. While he was clearly convinced that his museum without walls would take the process of intellectualisation to the extreme, his remarks leave no doubt in our minds that every piece in his collection has a referent in the world of objects. Malraux's museum without walls is both an inventory and a phenomenology. In the years since André Malraux conceived his museum and opened it to the public through his publications, reproduction technology has been revolutionised by the new media. Technical advances in image processing have made it possible to create pictures which look like reproductions but are actually outright inventions. The mode of reproduction used in the art of Susanne Fankhauser shows that the artist assigns the original and the reproduction to different categories of order. Her museum belongs to the world of images. The referent of a processed image is the reproduction and not its source. Reproduction is used not only to represent a reality that lies outside the visual image but to create an imaginary place as well.

Die Pose

Sarah Rossiter

Die Photographie ist nicht nur eine verhältnismäßig neue Erfindung;
verglichen mit dem Zeitraum, den unsere Kulturgeschichte umspannt,
ist sie vor allem auch eine sehr junge Kunst. Während Malerei und Bild-
hauerei mit der traditionell handwerklichen Verankerung auch die gesamt-
gesellschaftlichen Funktionen einbüßten, bildet die Gebrauchsphotogra-
phie für die freie Photographie nicht nur das geschichtliche Fundament,
sondern als immer noch praktizierte Disziplin bis heute eine zentrale Referenz. Am Anfang der Photographiegeschichte steht das Versprechen, das
Sichtbare getreuer abzubilden als jede andere Kunst vor ihr. Von vielen
Modernen ist bekannt, daß sie sich mit der Photographie praktisch und
theoretisch beschäftigten und in ihrer täglichen Arbeit im Atelier photographische Bilder als Bildvorlagen, Stimulans oder Skizzenersatz heran-
zogen. Die Photographie ist aber nicht, wie vielfach behauptet wird, an die
Stelle der Malerei getreten, sondern sie hat lediglich die Malerei von Teil-
aufgaben entbunden und um neue Möglichkeiten der Darstellung berei-
chert. Mit dem Auftauchen der ersten Photographien veränderte sich die
visuelle Auffassung der Welt, wovon wiederum die Kunst nicht unbeein-
flußt blieb. Umgekehrt vermochte kein künstlerisches Medium eine Bedeutung für das Selbstverständnis der Menschen zu erlangen, die auch nur
annähernd mit derjenigen der Photographie, insbesondere des Bildnisses,
vergleichbar wäre.

=

Sarah Rossiter wurde 1970 in Ithaca an der amerikanischen Ostküste
geboren, in eine Kultur, in der der Repräsentation des Individuums
durch die Photographie und der durch sie vollzogenen Einbindung des Einzelnen in die Familie, symbolisch dargestellt in den kleinen gerahmten
Photographien lebender und verstorbener Angehöriger auf dem Schreib-
tisch amerikanischer Präsidenten, fast kultische Bedeutung zukommt.
Sarah Rossiter ist als Plastikerin und Fotokünstlerin sowie als freie Aus-
stellungskuratorin tätig. Das Engagement für andere Künstlerinnen ist
eine Konsequenz ihres Selbstverständnisses als Künstlerin. In Bern zeigte
Sarah Rossiter eine Serie von Selbstbildnissen, in der sie sich abgebildet
hat, während Lichtbilder nach Gemälden ihrer Mutter Ann Kinner auf
ihren eigenen Körper projiziert wurden. In jenem Jahr, in dem Sarah
Rossiter geboren wurde, studierte ihre Mutter an der Cornell University
Malerei. Vater und Mutter von Sarah Rossiter waren bei ihrer Geburt
18 Jahre alt. Die Eltern trennten sich zwei Jahre später. Sarah Rossiter
lebte nun beim Vater, damit ihre Mutter das Kunststudium fortsetzen und

abschließen konnte. Ann Kinner setzte sich in den frühen siebziger Jahren
vor allem mit der amerikanischen Farbfeldmalerei auseinander. 1973 entstand eine Reihe von Arbeiten auf Papier, auf die sich Sarah Rossiter in
ihren eigenen Werken bezieht. Sarah Rossiter spricht von den frühen siebziger Jahren als den besten im künstlerischen Leben ihrer Mutter. Diese
hatte das gesamte zuvor entstandene malerische Werk zerstört und ver-
brannte bis auf 36 Arbeiten auch alle Gemälde aus dem Jahre 1973. In den
späten siebziger Jahren gab sie die ernsthafte Beschäftigung mit der Malerei auf. Sarah Rossiter kam mit diesen wenigen erhaltenen Arbeiten ihrer
Mutter erst in Berührung, als sie selbst schon als Künstlerin tätig war. Rossiter stellte bei sich selbst ein wachsendes Interesse für deren Werk fest.
Während sie die Gemälde photographisch dokumentierte, begann ein Prozeß der Annäherung an diese Malerei und der Auseinandersetzung mit
der Geschichte ihrer Mutter als Künstlerin, der in die vorliegenden Photographien mündete.

=

Sarah Rossiter konfrontiert sich in diesen Arbeiten als junge Frau und
Künstlerin mit ihrer Mutter als junger Malerin. Sie versucht, im Bild
die historische Ungleichzeitigkeit als Gleichzeitigkeit erscheinen zu lassen.
Sarah Rossiter untersucht ihr Verhältnis zu dieser Geschichte, die diejenige
ihrer Mutter und zugleich ihre eigene ist. Auch sie ist eine junge Frau am
Anfang einer künstlerischen Karriere, deren Verlauf unbestimmt ist. Sarah
Rossiter thematisiert ihre Rolle und Funktion im Leben ihrer Mutter, aber
auch deren Einfluß auf ihr eigenes Selbstverständnis als Frau und Künstlerin. Die Arbeiten fragen nicht zuletzt nach Kontinuität und Bruch in
weiblicher Produktion. Auf die Frage, weshalb sie sich auf einigen Arbeiten nackt zeige, antwortete Rossiter, sie beziehe sich auch damit auf Kind-
heitserinnerungen. Sich mit entblößtem Oberkörper öffentlich zu zeigen,
sei für ihren Vater selbstverständlich gewesen. Da sie ihre Kindheit ohne
Mutter beim Vater verbracht habe, seien junge Männer in ihrem Leben die
ersten Vorbilder für Unabhängigkeit, Überlegenheit und Rebellion gewesen. Das Bild des jungen Mannes ist positiv besetzt und als solches auf
ihr eigenes Geschlecht übertragen. Das Auge trifft denn auch nicht nur auf
einen nackten weiblichen Körper, sondern auch auf einen spezifischen
Blick und in Verbindung mit Waffen und Kleidung einer japanischen
Kampfsportart auf bestimmte Posen, die es dem Betrachter verunmög-
lichen, die Dargestellte zum Objekt zu machen. Verführung und Zurück-
weisung gehen Hand in Hand. Photographien wie diese wären in der
amerikanischen Kunst ohne Cindy Sherman nicht denkbar. Shermans
Photographien aus den frühen achtziger Jahren von Frauen in ausweg-
losen Situationen wirken wie Filmbilder und sind gleichzeitig genaue Nachstellungen von realen Momenten der Angst. Seither hat Sherman den Grad
der Inszenierung von Angst, Perversion, Gewalt und Schrecken kontinuier-

lich so weit gesteigert, daß sie ihre Arbeit, die inzwischen auch den Film umfaßt, dem Vorwurf aussetzt, die Evokationen des Schrecklichen seien ästhetischer Selbstzweck. Cindy Sherman thematisiert in ihren Arbeiten, was Photographen seit jeher wissen: Das Bildnis zeigt immer eine Pose. Wir sehen dasjenige Bild der abgebildeten Person, auf das sich diese mit dem Photographen während der Sitzung geeinigt hat. Rossiters Photographien funktionieren andererseits in vieler Hinsicht wie die Gemälde, ohne die ihre Arbeiten nicht entstanden wären. Der Minimalismus, die Farbe und die Sprache der Abstraktion waren Ann Kinners Weg, Identität und Gefühl zu formulieren. Rossiter versucht die Gemälde zu aktualisieren, indem sie durch ihre eigenen künstlerischen Mittel mit dem Werk ihrer Mutter interagiert. Der bildnerische Prozeß, der zu den Photographien führte, war eine Performance. Auf den Photographien bleibt nachvollziehbar, daß es sich bei den photographierten Gemälden um Projektionen handelt, denen die Künstlerin mittels Körperhaltung und Gestik ausdrucksvoll entgegentritt. Gleichzeitig versucht Rossiter, ihre eigenen Arbeiten denjenigen ihrer Mutter anzunähern, indem sie die Photographien in Rahmung und Präsentation wie Gemälde behandelt.

=

Die Arbeiten sind konzeptuell und bildhaft. Rossiter zeigt einen Prozeß der Überlagerung und Durchdringung von Emotionen, Geschichte und Geschichten, Raum, Farbe, Körper und Bewegung, ohne daß die einzelnen Elemente ineinander aufgehen. Vor allem aber zeigt sie die dargestellte Figur im Prozeß der Selbstbehauptung. In diesem entschlossenen Entgegentreten werden ihre Posen formel- und die Arbeiten bildhaft. Es ist diese ikonische Wirkung der Arbeiten, die mich auf Rossiter aufmerksam werden ließ und damit auch die Voraussetzung bildete für die Auseinandersetzung mit der Bilderzählung.

The Pose

Sarah Rossiter

Viewed against the background of our cultural history, photography is not only a relatively recent invention but a very young art as well. Whereas painting and sculpture lost both their traditional roots in manual craftsmanship and their broader social functions, utilitarian photography not only provided the historical foundation for artistic photography but, as discipline still practised in our time, continues to serve as its central point of reference. Early in its history, photography appeared to promise a more authentic representation of the visible world than had been achieved by any other art form before it. We know that many modern artists became involved with the theory and practice of photography and often used photographic images as sources, inspirational stimulus or substitutes for sketches in their daily studio work. Yet photography has not, despite frequent claims to the contrary, replaced painting. Instead it has simply liberated painting from some of its functions while enriching it with new possibilities for representation. The appearance of the first photographs changed our visual experience of the world, and art has not been immune to that influence. At the same time, no other artistic medium has had an impact on the human self-image that even approaches that achieved by photography, and particularly by the photographic portrait.

=

Sarah Rossiter was born in Ithaca, New York in 1970. Her roots are anchored in a culture in which the representation of the individual in photography and the integration of the individual within the family through photography, as expressed symbolically in the small, framed photos of living and deceased relatives on the desks of American presidents, has assumed almost religious meaning. Sarah Rossiter is a sculptress, a photo artist and an independent exhibition curator. Her commitment on behalf of other women artists is a logical consequence of her understanding of her own role as woman artist. In Bern, she exhibited a series of self-portraits made by projecting slide images of paintings by her mother Ann Kinner onto her own body. Sarah Rossiter's mother was a student of painting at Cornell University the year she was born. Both her mother and father were eighteen years old at the time of her birth. Her parents separated two years later, and Sarah lived with her father so that her mother could continue and complete her education. During the early 1970s, Ann Kinner was concerned primarily with American Color Field Painting. She completed a number of works on paper in 1973, to which Sarah Rossiter refers in her own work. Sarah Rossiter speaks of the early

1970s as the best years of her mother's life as an artist. Her mother later destroyed all of her previous work and burned all but 36 of the paintings she had done in 1973. She abandoned serious painting in the late 1970s. Sarah Rossiter did not become acquainted with these few surviving works of her mother's until she had already embarked upon her own career as an artist. From then on, she developed a growing interest in her mother's work. She documented the paintings in photographs and began an investigation into her mother's painting and her life as an artist, a process which produced the photographs featured in this book.

=

Sarah Rossiter confronts herself in these works as a young woman and artist with her mother as a young painter. She attempts to present the historical disparity as synchrony in her photographs. Sarah Rossiter examines her own relationship to this history, which is both her mother's and her own. She, too, is a young woman embarking upon a career in art, uncertain of the course that career will take. Sarah Rossiter examines her role and her place in her mother's life and their influence upon her own self-image as a woman and an artist. The photographs pose questions about continuity and disruption in female productivity. Asked why she appears nude in several of her works, Rossiter cites childhood memories. Her father found it perfectly natural to show himself bare-chested in public. Because she spent her motherless childhood with her father, young men were the first models of independence, superiority and rebellion in her life. Her image of young men is positive, and she has transposed it to her own gender. We see in these photos not only a semi-nude female body but also, in association with weapons and clothing representing a Japanese form of martial art, a specific view of certain poses that make it impossible for the viewer to see the portrayed subject as an object. Seduction and rejection go hand in hand. Photographs of this kind would be inconceivable in American art without Cindy Sherman. Sherman's photos of women in hopeless situations, done in the early 1980s, have the look of movie stills and are, at the same time, precise re-enactments of real moments of fear. Since that time, Sherman has continuously intensified the anxiety, perversion, violence and horror of her staged scenes to the point that she has been accused of pursuing in her work, which now involves film as well, horror effects as an end in themselves. What Cindy Sherman exposes in her works is a fact of which photographers have always been aware – that the portrait always exhibits a pose. The image we see of the portrayed subject is the image the subject and the photographer have agreed upon during the sitting. Rossiter's photographs, on the other hand, work in many ways like the paintings without which they would never have come to be. Minimalism, colour and the language of abstraction were the means with which Ann Kinner articulated her identity and her feelings. Rossiter seeks to give new currency to the paintings by interacting with her mother's art with the aid of her own artistic resources. The creative process that produced her photographs was a performance. In the photos, we recognise the photographed paintings as projections the artist confronts expressively through posture and gesture. At the same time, Rossiter seeks to close the distance between her own works and those of her mother by presenting and framing her photographs like paintings.

=

These works are both conceptual and visual. Rossiter shows us a process in which emotions, history, stories, space, colour, body and motion overlap and interpenetrate one another yet still retain their separate identities. Above all, however, she presents a view of the portrayed figure in the process of asserting itself. In this determined confrontation, its poses assume the quality of conceptual formulas, while the works themselves are visual. It is precisely this iconic effect evoked by her works that attracted my attention to Sarah Rossiter and formed the point of departure for my study of the visual narrative.

Cyborgs

Lee Bul

Wenn man von Cyborgs spricht, ist heute von einem Körper die Rede,
einem biologischen Replikanten, und nicht mehr nur von einer lite-
rarischen Figur wie jener aus dem im Jahre 1818 in Genf von der englischen Schriftstellerin Mary Shelley verfaßten Roman *Frankenstein oder
Der moderne Prometheus.* In ihrem Buch läßt die Autorin ihre Titelfigur
Dr. Frankenstein ein Lebewesen künstlich erzeugen. Zahlreiche Filme mit
dem Monster in der Titelrolle gehen auf diesen englischen Roman aus der
Zeit der industriellen Revolution zurück.

=

Michel Carrouges beschrieb 1954 in seinem Buch *Les Machines Célibataires* für die Künste das Entstehen eines neuen Mythos, dessen
Spur er bis ins 19. Jahrhundert zurückverfolgte. Den Titel übernahm er
von Marcel Duchamp, der ihn erfunden und in seinem *Großen Glas* visua-
lisiert hatte. Die Bedingung für die Vorstellung von Fortpflanzung als Technik der Reproduktion war der industrielle Fortschritt, als dessen Folge die
Maschine zum Sinnbild für den perfekten Menschen werden konnte. Während sich der Mensch aus der Retorte 1818 noch als Monster entpuppte,
war die Kopie, die Villiers de l'Isle-Adam 1886 in *L'Eve Future* beschrieb,
bereits perfekter als das Original. Unvergessen ist auch Ridley Scotts Film
Blade Runner aus dem Jahre 1982, der auf Philip K. Dicks Roman *Do
Androids Dream of Electric Sheep?* basiert. Der Film spielt in Los Angeles
im Jahre 2019 und erzählt die sich entwickelnde, verbotene Beziehung
zwischen Rick Deckard, der für Geld künstlich erzeugte Lebewesen ver-
folgt und tötet, und Rachael, einer Frau aus der Produktion von Eldon
Tyrell, der im Unterschied zu Frankenstein nicht aus wissenschaftlichem
Interesse, sondern aus kommerziellen Gründen handelt. Seine Erzeugnisse
sind so perfekte Kopien, daß sie kaum von Menschen zu unterscheiden
sind. Es gibt im Film nur eine einzige sichere Methode, einen Replikanten
zu erkennen, die Prüfung der Pupille, die keine Gefühlsregung zu weiten
vermag. Der Film setzt eine gegenläufige Bewegung in Gang. Während Rick
Deckard, bis er sich in Rachael verliebt, seine Menschlichkeit zusehends zu
verlieren scheint, fordern die künstlich erzeugten, seelenlosen Kreaturen
diese für sich ein. Rachael ist die einzige Figur im Film, die nicht zu wissen
scheint, wer sie ist. Sie imitiert die Umgebung, in der sie lebt, und handelt
gemäß dem Kontext, in dem sie sich bewegt, bis sich in ihr der Verdacht
regt, kein Mensch, sondern eine Maschine zu sein. Die Selbsterkenntnis
macht sie menschlicher und hilft Deckard, zu seinem eigenen Menschsein
zurückzufinden. Rachael ist nach heutigem Sprachgebrauch ein »Cyborg«.

Der Begriff stammt aus der Welt der Sciencefiction und ist die Abkürzung für »cybernetic organism«. Er bezeichnet nach Auffassung der amerikanischen Wissenschaftshistorikerin Donna Haraway allerdings nicht nur Hybride, künstlich erzeugte technologisch-organische Wesen, sondern meint alle in einer postmodernen Welt lebenden Menschen.

=

Ob in Film, Theorie oder Kulturgeschichtsschreibung, die Bezeichnung »Cyborg« meint in jedem Fall eine Erzähl- oder Argumentationsfigur und nicht eine Realität wie die, mit der uns die entfesselte Kunst der Biotechnologie seit wenigen Jahren konfrontiert. Das Klonen von Säugetieren ist bereits gelungen, die Herstellung menschlicher Gewebe steht kurz vor der Verwirklichung. Die Welt der Kunst dagegen ist diejenige der Darstellung: die Übersetzung von Erfahrungen, Beobachtungen und Ideen in bildhafte Formeln. Zwar wurde der Kunstbegriff im 20. Jahrhundert stark erweitert – die Verbindung von Kunst und Leben ist denn auch bis heute ein immer wiederkehrendes Thema künstlerischer Praxis und Theorie –, doch haben die Künstler trotz der ständigen Ausweitung ihres Handlungsspielraums nur ausnahmsweise die Biologie des Körpers in ihre Arbeit miteinbezogen. Die Manipulation des Lebens blieb mit gutem Grund meist tabuisiert. Der amerikanische Künstler Forrest Bess (1911–1977) allerdings verstand seine Kunst als inneren Auftrag, dem er wie ein naturwissenschaftlicher Forscher nachlebte, mit einer Unerbittlichkeit, der kein Preis, nicht einmal die Selbstverstümmelung, zu hoch war. Er bezeichnete sich selbst nicht als Maler, sondern als Alchimist. Er begann im Jahre 1934 nach einem Architekturstudium zu malen. In Bay City, Texas, eröffnete er im Jahre 1947 ein Anglergeschäft und nahm die während der Kriegsjahre unterbrochene Malerei wieder auf. Ab 1953 korrespondierte er mit Psychiatern und Anthropologen über seine Theorie der Androgynie als Idealzustand des Menschen. Er sammelte Unterlagen zur Zweigeschlechtlichkeit in Mythologie und Religion und entwickelte in seiner Kunst eine entsprechende Symbolik, ließ es jedoch nicht bei dieser elaborierten Bildsprache bewenden. Im Jahre 1960 schnitt sich der Künstler unterhalb seines Penis eine Öffnung. Zahlreiche Operationen folgten. Forrest Bess versprach sich von seiner Umwandlung in einen Hermaphroditen, die ihm zusätzlich zu seiner männlichen Sexualität die Erfahrung des weiblichen Orgasmus ermöglichen sollte, das Aufhalten des Alterns und psychische Ausgeglichenheit.

=

Forrest Bess orientierte sich an einer zu verwirklichenden Vision, Lee Bul dagegen bezieht sich auf einen Diskurs. Ihre *Cyborgs* sind den in Korea sehr populären weiblichen Cyborgs aus Animation und Comic nachempfunden. Cyborgs sind in den japanischen und koreanischen Comics unbezwingbare, unerschrocken und kaltblütig handelnde Wesen mit ein-

deutig weiblichen Geschlechtsmerkmalen und mädchenhaften Gesichtern.
Lee Bul erwähnt als eine für sie besonders aufschlußreiche Eigenschaft der
Cyborgs, daß diese meistens durch einen Mann oder Jüngling program-
miert und kontrolliert werden. Die Paarung von übermenschlicher Kraft,
Güte und emotionaler Verletzlichkeit bezeichnet die Künstlerin als klassi-
sche Männerphantasie. Sie arbeite mit der Figur des Cyborgs, um zu zeigen, daß bestimmte Vorstellungen – etwa aus der Geschlechtergeschichte
– durch den wissenschaftlich-technologischen Fortschritt nicht überwun-
den werden. Zwar orientierte sie sich bei der Erarbeitung der in Bern ausgestellten Werkgruppe an einer vor allem unter Jugendlichen populären
Bildsprache, doch die Rolle der Massenmedien in unserer Gesellschaft ist
nicht das Thema der *Cyborgs*. Thema ist vielmehr das Menschenbild in
hochindustrialisierten Gesellschaften des ausgehenden 20. Jahrhunderts.
Lee Bul verwendete Silikon, um die Außenhaut der *Cyborgs* herzustellen,
ein Material, das bekanntlich in der Medizin, insbesondere in der Schön-
heitschirurgie, dazu dient, Implantate aufzubauen. Im Unterschied zu den
verführerischen und technisch perfekten massenmedialen Vorbildern wirk-
ten die handwerklich unperfekt gefertigten, kopflosen, einarmigen und ein-
beinigen Figuren sowohl in ihrem makellosen Weiß als auch in ihrem frag-
mentarischen, gepanzerten Zustand wie Gipsskulpturen der klassischen
Kunst. Wie Marionetten hingen die vier Plastiken an langen Stahlseilen von
der Decke des Ausstellungssaals.

=

Das Spannungsfeld, in dem die Künstlerin arbeitet, ist jenes zwischen
Natur, Kultur und den durch Wissenschaft und Technik geschaffenen
materiellen und sozialen Lebensbedingungen. Lee Buls neue Plastiken sind
Hybride, die nicht nur an die Welt der Sciencefiction und die entfesselte
Kunst der Biotechnologie erinnern, sondern welche auch an den Schöp-
fungsmythos, die älteste Erzählung und den ersten Bericht eines künstlerischen Aktes überhaupt, denken lassen. Zwar ist der Cyborg eine Vorstellung aus der Zukunftsforschung, doch hinter der Idee des Cyborgs
verbirgt sich jene der Perfektion und damit eine der ältesten Obsessionen
der Menschheit. Die Idee der Unsterblichkeit und der Wunsch, Lebewesen
nicht durch biologische Zeugung, sondern durch eine des Geistes, eine der
Kunst zu schaffen, hat den Menschen durch die Geschichte begleitet und
wird von Lee Bul in diesen *Cyborgs* kritisch aufgegriffen. Faszinierend am
bisherigen Werk von Lee Bul ist, wie die Künstlerin in kurzer Zeit von der
direkten Zurschaustellung des Körpers, beispielsweise in Performances
wie *Abortion* (1989) oder *Sorry for Suffering* (1990), zur Darstellung des
künstlichen Körpers gelangte, ohne ihren emanzipatorisch-aktivistischen
Standpunkt aufzugeben. Lee Bul baumelte in jenen frühen Arbeiten kopf-
über und nackt über dem Publikum, berichtete von einer Abtreibung und
rezitierte Gedichte und Pop-Songs, und sie zeigte sich in der Öffentlichkeit

mit weichen, inneren Organen nachgebildeten Skulpturen, die ihren Körper wie ein Kleid umhüllten. Die *Cyborgs* von Lee Bul sind Erkenntnisinstrumente, weil sie die Schnittstelle zwischen Kunst, Biotechnologie und feministischer Sozialwissenschaft anzeigen, dabei aber nicht die Frage nach der Wahrheit dieser Phantasie beantworten, sondern diejenige nach ihrer Herkunft aufwerfen.

Susanne Fankhauser
Das Museum der Tiere, 1998

Susanne Fankhauser
Das Museum der Tiere, 1998

66

Sarah Rossiter
Velocity (Grid with Sticks), 1998, *Velocity (Orange Stripe),* 1998

68

Sarah Rossiter
Velocity (Grid with Sticks), 1998

Sarah Rossiter
Velocity (Orange Stripe), 1998, *Velocity (Drips with Nunchucks)*, 1998

Velocity (Yellow and Red Drips), 1998, *Velocity (Brown Drips),* 1998

Lee Bul
Cyborg W2, 1998

Lee Bul
Cyborgs W3, W4, 1998

Lee Bul
Cyborgs W1, W4, W3, 1998

74

Lee Bul
Cyborgs W1, W4, W3, 1998

Martina Klein
Ohne Titel (gelb hell/gelb hell), 1999, *Ohne Titel (braun dunkel/rot),* 1999

Martina Klein
Ohne Titel (gelb hell/gelb hell), 1999, *Ohne Titel (grün hell/rot dunkel)*, 1998

78

Ohne Titel (weiss/schwarz), 1998, *Ohne Titel (ocker hell/orange)*, 1998

Martina Klein
Ohne Titel (braun/beige), 1999
Ohne Titel (gelb hell/gelb hell), 1999
Ohne Titel (braun dunkel/rot), 1999

80

Cyborgs

Lee Bul

When we speak of cyborgs today, we are no longer talking about a body, a biological replication, or of a literary character of the type created by the English author Mary Shelley in her novel Frankenstein or the Modern Prometheus *(Geneva, 1818). Shelley's protagonist Dr. Frankenstein brings a man-made being to life. A number of movies featuring the monster in the title role are based on this English novel from the era of the Industrial Revolution.*

=

In his book Les Machines Célibataires, *published in 1954, Michel Carrouges described the emergence of a new myth, whose origins he traced back to the 19ᵗʰ century. Carrouges adopted the title from Marcel Duchamp, who had coined and rendered it in visual terms in his* Large Glass. *The concept of biological reproduction as a technical reproductive process was based upon the phenomenon of industrial progress, through which the machine, as its product, came to symbolise the perfect human being. While the man-made human being turned out to be a monster in 1818, the clone described in 1886 by Villiers de l'Isle-Adam in* L'Eve Future *was more perfect than the original. Also unforgettable is Ridley Scott's* Blade Runner *(1982), a cinematic adaptation of Philip K. Dick's novel* Do Androids Dream of Electric Sheep?. *Set in Los Angeles in the year 2019, the movie tells the story of the forbidden relationship that develops between Rick Deckard, a hired hunter and killer of androids, and Rachael, a woman produced in the laboratory of Eldon Tyrell, a man whose work, unlike Dr. Frankenstein's, is not motivated by scientific interest but by purely commercial considerations. His creations are such perfect copies that they are virtually impossible to distinguish from real human beings. The film suggests only one reliable method for detecting a replication: observation of the pupil, which is incapable of dilating in response to emotional stimulus. The plot sets opposing currents in motion. While we see Rick Deckard progressively losing his human qualities, until the point at which he falls in love with Rachael, the artificial, soulless creatures seem to grow more and more human as time goes by. Rachael is the only character in the film who does not appear to know who she is. She adapts to her surroundings und behaves in keeping with whatever context she finds herself in. Eventually, however, she begins to suspect that she is not a human being but a machine. Self-awareness makes her more human and helps Deckard rediscover his own humanity as well. Rachael is a "cyborg" in the contemporary sense of the word. The term comes from Science Fiction and is an abbrevia-*

tion for "cybernetic organism." According to Donna Haraway, an American historian of science, it denotes not only hybrids – artificially created techno-organic beings – but actually stands for all human beings who live in a post-modern world.

=

Regardless of whether it is used in film, theory or cultural historiography, the term "cyborg" refers to a figure in narrative or argumentation and not to a real being of the kind with which the unfettered art of bioengineering confronts us. Mammals have already been successfully cloned, and the production of human tissue is just around the corner. The world of art is the world of representation – the translation of experiences, observation and ideas into visual formulas. The concept of what art is has been substantially expanded in the course of the 20th century, and the link between art and life remains a recurring theme in both the theory and practice of art. Yet though artists enjoy ever-greater freedom of action, they have only rarely made the biology of the human body a focus in their work. For good reasons, the manipulation of life has remained for the most part taboo. The American artist Forrest Bess (1911–1977) saw his art as an inner mandate to which he responded with the commitment of a research scientist and a merciless dedication for which no price, not even that of self-mutilation, was too high. He referred to himself as an alchemist rather than a painter. Bess started painting in 1934 after completing his studies in architecture. In 1947 he opened a bait and tackle shop in Bay City, Texas and took up his paints again for the first time since the Second World War. He began corresponding with psychiatrists and anthropologists on the subject of his theory of androgyny as the ideal human state in 1953. Bess compiled material on androgyny from religious and mythological sources, developing at the same time a corresponding system of symbols in his art. But he was unable to rest content with this elaborate language of images alone. In 1960 the artist made an incision in his body below his penis, and numerous other operations followed. Forrest Bess hoped that by transforming himself into a hermaphrodite, and thus enabling himself to experience both male sexual pleasure and female orgasm, he could hold back the ageing process and achieve a state of psychic balance.

=

Forrest Bess focused upon a vision and attempted to realise it. Lee Bul, on the other hand, refers to a pattern of discourse. Her Cyborgs are adapted from popular female cyborg figures in comics and cartoons. The cyborgs who populate Japanese and Korean comics are fearless, invincible, cold-blooded beings with unmistakably female sexual features and girlish faces. Lee Bul mentions the fact that these comic cyborgs are ordinarily programmed and controlled by a man or a male adolescent as one of their most revealing characteristics. The artist sees the union of super-

human strength, technical prowess, virtue and emotional vulnerability as a classical male fantasy. She uses the cyborg figure to demonstrate that certain notions – including products of gender history – have not been erased through progress in science and technology. Although she drew upon visual imagery popular among today's youth in creating the group of works exhibited in Bern, her Cyborgs are not concerned with the subject of mass media in our society. Her theme was the image of the human being in the advanced industrial societies of the waning 20th century. Lee Bul made the skin of her Cyborgs out of silicone, a material known to have medical applications, particularly in cosmetic surgery, where it is used in implants. In contrast to the seductive and technically perfect models presented in the mass media, the immaculate whiteness and fragmentary, armoured appearance of her imperfect, hand-crafted, headless, one-armed and one-legged figures calls to mind classical plaster sculptures. The four figures were suspended like marionettes on long steel cables from the ceiling of the exhibition hall.

=

The artist focuses on the field of tension involving nature, culture and the materials and social conditions engendered by science and technology. Lee Bul's new sculptures are hybrids that not only remind us of the world of Science Fiction and the unfettered art of biotechnology but recall as well the myth of Creation, the oldest of all stories and the very first account of a creative act. Although the concept of the cyborg is a product of futurist research, the idea embodies the dream of perfection and thus one of mankind's oldest obsessions. The vision of immortality and the desire to create living beings as products of the mind, of art, rather than of biological reproduction, is a constant in human history. Lee Bul subjects it to critical examination in these Cyborgs. A fascinating aspect of Lee Bul's work to date is the artist's rapid progress from direct exhibition of the human body – in performances such as Abortion *(1989) or* Sorry for Suffering *(1990), for instance – to the presentation of artificial bodies, a transition she has accomplished without abandoning her emancipatory, activist point of view. In these earlier works, Lee Bul dangled upside down and naked above her audience, talking about her abortion and reciting poems and pop songs, and she exhibited herself in public with soft sculptures shaped like internal organs that enveloped her body like a dress. Lee Bul's Cyborgs are instruments of knowledge. They reveal the point at which art, biotechnology and feminist social science intersect. Yet they do not answer the question about the truth of this fantasy but inquire instead about its origin.*

Von der Farbe des Widerscheins

Martina Klein

Die Rückseite eines Gemäldes als ein für das Werk konstitutives und der Bildseite gleichgestelltes Element zu zeigen, hat vielfache Implikationen. Zunächst werden wir daran erinnert, daß Kunstwerke im konkreten wie im übertragenen Sinne sowohl offenbaren als auch verbergen. Kunstwerke werden auch als Artefakte bezeichnet, um ihr Gemachtsein zusätzlich zu betonen. Ein Blick in die Geschichte lehrt jedoch, daß die Arbeiten nur selten zu erkennen geben, unter welchen geistigen, sozialen und ökonomischen Voraussetzungen die Künstler gearbeitet haben. Aus der Anschauung erfahren wir weder Namen von Auftraggebern, noch weshalb ein Werk an einem bestimmten Ort ausgestellt oder aufbewahrt wird. Es gibt malerische Techniken, die für immer ein Geheimnis bleiben werden, und Bildinhalte, die dem Vergessen anheimgefallen sind und die deshalb erst wieder von Kunsthistorikern ermittelt und in Worte übersetzt werden müssen. Die Gleichstellung von Vorder- und Rückseite erinnert daran, daß seit dem 19. Jahrhundert bis in unsere Gegenwart eine Tendenz weg vom symbolischen Ausdruck hin zur Darstellung und Thematisierung von Strukturen festzustellen ist. Trotzdem gibt es weiterhin Malerei, die sich als Medium der Bildschöpfung zu legitimieren vermag. Luc Tuymans etwa ist ein Maler der Generation von Martina Klein, der überzeugend an einer Malerei der Bilder weiterarbeitet. Die Tätigkeit von Luc Tuymans als Maler ist diejenige eines Übersetzers. Seine Bilder erscheinen, als ob sie auf Dokumente verwiesen. Mit der Kritik der Bildmagie aber, deren Macht in der Werbung ungebrochen scheint, und ihrer bewußten Schwächung durch die Künstler des 20. Jahrhunderts ging die Stärkung des instrumentellen Charakters des Kunstwerks einher.

=

Die Arbeiten von Martina Klein beschreiben sowohl den Raum und den Ort der Malerei als auch den Raum und den Ort des Betrachters. Es ist von einer Malerei die Rede, die Bildraum als Farbraum zeigt und den Ort als Wirkung des Zusammenspiels von malerischer und architektonischer Ordnung begreift. Von einem Betrachter, der die Welt nicht als bewegtes Bild auf dem Bildschirm, als Fläche, sondern als strukturierte Umgebung erfährt, in der er sich als Körper vor, entlang, durch und zwischen anderen Elementen bewegt. Für diese Haltung bezeichnend ist, daß Martina Klein schon am Tag ihrer Ankunft in Bern auffiel, aus welchem Stein die Altstadt gebaut ist. Sie hat die Farbe des Sandsteins wahrgenommen. Sie hat auch festgestellt, daß die Dächer weiter als anderswo in die Gassen vorragen und dadurch einen Raum bilden, der mit demjenigen der

84

Laubengänge korrespondiert. Material, Farbwert, Zwischenraum und Perzeption sind Schlüsselbegriffe für das Verständnis ihrer künstlerischen Arbeit.

=

Martina Klein arbeitet seit fünf Jahren an zweiteiligen Bildern. Jedes Werk besteht aus zwei monochromen und quadratischen Flächen, die so miteinander verbunden sind, daß sie einen rechten Winkel bilden. Es ist dies das einfachste Verfahren, um von der Fläche in den Raum zu gelangen. Die beiden Bildhälften sind meistens verschiedenfarbig. Sichtbar ist bei der einen, rechtwinklig von der Wand weggedrehten Hälfte der Bildfläche die Vorder- und die Rückseite. Die Keilrahmen sind während des Farbauftrags abgeklebt und zeigen deshalb keine Spuren des Arbeitsprozesses, die Rückseiten verweisen nicht auf die Vorderseite. Die funktionale Bedeutung des Keilrahmens als Spannrahmen ist zwar ablesbar, hat aber wegen den fehlenden Arbeitsspuren keinen weiterführenden Stellenwert. Die mit dem Spachtel gleichförmig aufgetragene Ölfarbe unterstützt die faktische Wirkung der Gemälde. Es wird der Eindruck erzeugt, deren Vorder- und Rückseiten gehörten der gleichen Ordnung an. Eine Ausnahme bilden die kleinen Arbeiten, die schon als Winkel bemalt und nicht erst später zu Winkeln zusammengestellt wurden. Die beim Arbeitsprozeß auf den nicht abgeklebten Rückseiten entstandenen Farbspuren sind bei diesen kleinen Werken nicht entfernt worden. Die Dimensionen von Hand- und Bildfläche liegen bei diesen Arbeiten so nahe beieinander, daß die Arbeitsspuren als Zeichnungen erscheinen.

=

Die Plazierung der Arbeiten im Raum und bezogen auf weitere Gemälde fördert oder unterbindet die persönliche Bezugnahme durch das Publikum, schafft konfrontative Situationen oder setzt optische Wechselwirkungsprozesse zwischen den Farbräumen in Gang, deren Fülle sich allerdings nur demjenigen Betrachter mitteilt, der sich vor und in den Werken bewegt. Die kleinen Winkel bilden optisch begehbare Farbvolumen. Die großen Winkel sind wie architektonische Räume betretbar und durchschreitbar. Der Raumbezug der Arbeit von Klein erfuhr in dieser Ausstellung eine zusätzliche Akzentuierung, weil die erstmals zu sehenden Großformate nicht an der Wand hingen, sondern wie unvollendete Gemälde im Atelier auf drei Hölzchen frei im Raum standen. Mit diesen Arbeiten hätte Martina Klein den Schritt von der Malerei zur Architektur machen können. Zwar zeigte sie Malerei in installativer Form, doch betont wurden weiterhin auch die malerischen Aspekte ihrer Arbeit. Die Bildbetrachtung bedingt die Bewegung des Betrachters und der Betrachterin nicht nur, weil die Werke selbst Raum einnehmen, sondern auch, weil die Trennung in Schau- und Rückseite aufgehoben ist und damit auch die Fokussierung auf die Bildfläche. Man kann sich, schreibt die Künstlerin, in einer Ausstellung der zweiteiligen Arbeiten »normal weiterbewegen, kann gehen, stehen bleiben, sich hineinsetzen, kann sich dahinter stellen. Man kann sich selbst sehen. Es gibt nicht nur einen Blickwinkel. Für mich sind es keine Verstecke, sondern offene Räume, und so versuche ich sie auch zu installieren«.

=

Für den Projektraum erarbeitete Martina Klein eine Ausstellung, in der sie erstmals die ganze Raumhöhe in die Hängung einbezog und somit mit ihren Arbeiten unterschiedliche Ebenen im Raum anzeigte. Die Wechselwirkungsprozesse zwischen den Arbeiten erfuhren dadurch gegenüber der bislang üblichen Hängung der Gemälde auf Augenhöhe eine Dynamisierung. Martina Klein plazierte die Arbeiten so, daß diese möglichst offene, komplexe Beziehungen untereinander, aber auch zum Raum und sogar zur Raumeinrichtung eingingen. Erstmals diente eine Arbeit als Hängefläche für eine andere, womit der instrumentelle Charakter ihrer Werke zusätzlich unterstrichen wurde. Je nach Hängung eines Werks kann die eine monochrome Fläche zum Spiegel der andern werden, wird seitlich einfallendes Licht in leicht veränderter Tönung reflektiert oder können farbige Schattenzonen geschaffen werden. Die Rückseite, Schatten, das Verborgene sind mitgedacht und mitgesehen. Sie haben in dieser Arbeit denselben Stellenwert wie der Farbraum. Der für das einzelne Werk durch dessen Winkelform konstitutive Zwischenraum wurde in vielfacher Form auch Thema der Ausstellung. Martina Klein macht Raum erfahrbar als Zwischenraum. Der französische Philosoph Gaston Bachelard schreibt: »Jeder Winkel in einem Haus, jede Ecke in einem Zimmer, jeder eingezogene Raum, wo man sich gern verkriecht, sich in sich selbst zusammenzieht, ist für die Einbildungskraft eine Einsamkeit, der Keim eines Zimmers, der Keim eines Hauses.« Der Winkel ist aber auch, und darüber schreibt Bachelard ebenfalls, in Gegenwirkung zu dem im Winkel erlebten Zustand der Konzentration, der Enge und Unausweichlichkeit eine Aufforderung zum Aufbruch. Kein Winkel kann permanenter Ort des Daseins sein. Winkel erzeugen Bewegung.

The Colour of Reflection

Martina Klein

The presentation of the back of a painting as a constituent element of the work, equal in importance to the image-bearing side, has a number of implications. We are reminded above all that works of art not only reveal but conceal, in both a concrete and a figurative sense. Works of art are often referred to as artefacts, a term that emphasises their character as made objects. A look at history suggests, however, that artists' works only rarely reveal the mental, social and economic situations in which their authors worked. The viewing process tells us neither the name of the person who commissioned a work nor the reason why it is exhibited or preserved at a given location. Some painting techniques will remain mysteries forever; some contents have long been forgotten and must therefore be rediscovered and translated into words by art historians. The equation of the front and back of a painting serves as a reminder of the tendency to abandon symbolic expression in favour of the examination and representation of structures that emerged in the 19th century and has continued into our time. Yet a kind of painting that derives legitimacy from its function as a medium of image creation persists. Luc Tuymans is a painter of Martina Klein's generation who continues to paint pictures. His role as a painter is that of a translator. His paintings seem to refer to documents. Nevertheless, critique of the magic of images, whose power apparently lives on undiminished in advertising, and deliberate attempts of 20th-century artists to diminish it have gone hand in hand with a strengthening of the instrumental character of the work of art.

=

Martina Klein's works describe the space and place of the painting as well as the space and place of the viewer. It is a kind of painting that deals with pictorial space as colour space and place as the effect of the interplay of painterly and architectural orders. Viewers of such painting do not experience the world as a moving image on a screen, as a flat surface, but as a structured environment in which they moves as bodies along, through and among other elements. The fact that Martina Klein took note of the kind of stone used in the building of the old town centre in Bern the very day she arrived in the city is characteristic of her perceptual attitude. She noticed the colour of the sandstone. She also recognised that the roofs of the buildings protruded further out over the streets than in other towns, forming a space which corresponds to that of the famous arbours of Bern. Aspects of material, colour values, in-between space and perception are keys to an understanding of her art.

Martina Klein has been working on two-part paintings for about five years. Each work consists of two square, monochrome fields joined in such a way that they form a right angle. This is the simplest technique for moving from the flat plane to three-dimensional space. The two halves of the paintings are usually of different colours. Both the front and back sides of the half set at a right angle to the wall are visible. The canvas stretchers were taped during the application of the paint and thus show no traces of the work process. Although the functional significance of the stretcher is apparent, the absence of signs of the painting process renders it otherwise relatively unimportant. Uniformly applied with a pallet knife, the paint enhances the factual effect of the painting. The impression evoked is that of front and back sides of equal value. Exceptions to the rule are the small works, which are painted as angle configurations rather than assembled after painting. The traces of paint left on the exposed backs of the paintings during the work process have not been removed from these small pieces. The dimensions of the pictorial surface and the artist's hand are so similar in these works that the traces of the work process take on the appearance of drawn elements.

=

The placement of the paintings within the room and in relation to other paintings either fosters or hinders personal identification by viewers, creating confrontational situations or triggering interactive visual processes among the colour fields. The richness of these processes is evident only to viewers who move about in front of and within the works. The small angles form visually accessible colour spaces. The large angles can be entered and walked through, like architectural volumes. The spatial orientation of Klein's works was additionally accented at this exhibition, as the large-scale pieces (exhibited here for the first time) were not hung on the wall but left standing like unfinished works on three blocks of wood. Martina Klein could well have made the transition from painting to architecture with these works. She presented painting in installation form, yet the painterly aspects of her work were emphasised as well. The viewing process influenced the movement of viewers within the room in two ways: The works themselves occupied space, and the distinction between view and back sides was eliminated, thus preventing concentration on the pictorial surface alone. It is possible, the artist writes, "to move about 'normally' in an exhibition of the two-part paintings. Viewers can walk, stop and stand, sit down inside the works or stand behind them. They can see themselves. There is more than one point of view. I don't see [the works] as places to hide but as open spaces, and I try to install them that way".

=

Martina Klein developed an exhibition for the "Projektraum" in which she made use of the full height of the wall in the hanging scheme, thus making it possible to present her works at different levels within the room. In this way, she was able to imbue the interactions among the works with a dynamic quality that could not have been achieved through the usual practice of hanging the paintings at eye level. She positioned the works in such a way that they could form relationships of maximum openness and complexity with one another, with the surrounding space and even with the room furnishings. For the first time, one painting served as a hanging surface for another, thus further emphasising the instrumental character of her works. Depending upon the manner in which it is hung, the monochrome surface of a given work can become a mirror of another; light striking a painting from one side is reflected in a slightly modified hue and zones of coloured shadow may appear. The backs of the paintings, shadows and concealed aspects are taken into account and exposed to view. In these works, they are just as important as the painted surfaces. In many different ways, the in-between space formed within the individual work by virtue of its angular configuration became a theme of the exhibition. Martina Klein makes space perceptible as in-between space. The French philosopher Gaston Bachelard writes: "Every corner of a house, every corner of a room, every encompassed space where one takes refuge, gathering oneself together inside it, is a form of solitude for the imagination, the germinal cell of a room, the seed of a house." Yet in contrasts to the state of concentration, of confinement and inescapability experienced in the corner, the corner is also – as Bachelard points out as well – a call for change. No corner can be a permanent place of existence. Corners generate movement.

Arbeit im Netzwerk

@ home

@home ist die Bezeichnung für einen Ort, eine Homepage und das Künstlerkollektiv, das in wechselnder Zusammensetzung diesen Ort in Basel seit 1995 bearbeitet, bewohnt und bespielt. Der Ort ist ein altes Gebäude, das eine wechselvolle Geschichte hinter sich hat und in absehbarer Zeit abgerissen werden soll. Das vom Künstlerkollektiv inzwischen für seine Bedürfnisse umgebaute Haus sowie die umliegenden Gebäude dienten vor dem Ersten Weltkrieg als Durchgangsstation für Auswanderer aus dem Balkan, die über die Schweiz in die Vereinigten Staaten reisen wollten. Von dieser Bestimmung der architektonischen Anlage erzählt ein von einem Mitglied der damaligen Eigentümerfamilie verfaßter Roman, der, so die Gruppe, vor vier Jahren im verlassenen Gebäude gefunden wurde. Dieser Fund wurde als Bestätigung der Raumwahl aufgefaßt, zeigte das Buch doch eine Kontinuität in der Funktion des Hauses unter anderen historischen und gesellschaftlichen Bedingungen an. Das folgende Statement findet sich in einem Arbeitspapier der Gruppe: »Die letzten Jahre verdeutlichen unwiderruflich die Veränderungen des Aggregatzustandes unserer Kultur. In der Erscheinung zeigt sich dieser Aspekt in der zunehmenden globalen Nomadisierung der Kultur wie im Erwachen des Bewußtseins für ein holotropes Nebeneinander von sich durchdringenden Realitäten sowie der Auflösung der Idee einer linearen Ursächlichkeit.«
=
@home hat eine Geschichte, die bis in die frühen achtziger Jahre zurückreicht. Als Folge der mit ihren Anliegen in der Schweizer Öffentlichkeit gescheiterten Jugendbewegung der achtziger Jahre, deren politisch aktive Protagonisten in besetzten Häusern lebten, entstanden auch in Basel an verschiedenen Orten Kollektive, die ihren Alltag und insbesondere ihren Lebensort zum Thema ihres Zusammenlebens machten. Politische Agitation wurde abgelöst durch eine bewußt künstlerische und selbstverantwortliche Lebenspraxis, die ideell an Joseph Beuys und dessen Erweiterung des Kunstbegriffs zur sozialen Plastik anschließt. Als erste Arbeit führt @home im Werkverzeichnis den Um- und Ausbau des Hauses und dessen Einrichtung mit Mobiliar auf Rädern. Beweglichkeit, Spontanität, Veränderbarkeit und Präsenz sind einige Schlüsselbegriffe für das Lebens- und Kunstverständnis dieser jungen Künstlerinnen und Künstler. Es gibt kaum Produkte aus dem Atelier @home, aber zahlreiche Produktionen wie beispielsweise das Happening *Music Circus* nach John Cage, das im Gründungsjahr stattfand. Situationsbezogene Environments sind die wichtigste Ausdrucksform des Kollektivs. @home veranstaltete 1996 und

1997 Pop-Events, die als *Hometrainerclub* bekannt und von diversen DJs, Musikern und Künstlern aus dem In- und Ausland besucht wurden. 1998 zeigte @home die *Flugschule by transphere®*, eine Installation mit ferngesteuerten Flugobjekten, Licht- und Diaprojektionen auf schwebende Kugeln, interaktivem Soundsystem und Hare-Krishna-Essen. 1999 organisierte die Gruppe eine *Mondmilchfahrt* an den Lac des Taillères (CH). Die Veranstaltung umfaßte eine Carfahrt, ein Outdoor-Happening im Schnee und eine Karaoke-Party. @home verbindet den Freiheitsbegriff von Joseph Beuys, den dieser 1978 als »positiven Anarchiebegriff« definierte, mit dem Zeitgeist der neunziger Jahre, der in Mode und Design ein Revival der siebziger Jahre brachte. @home zeigt Privates öffentlich, macht Kollektives im Individuellen erkennbar und verbindet Persönliches mit Gesellschaftlichem. Unverkennbar ist, daß es sich bei den Arbeiten der Gruppe um Inszenierungen handelt. Betont wird von den Künstlerinnen und Künstlern insbesondere, daß @home einen bestimmten Ort meint, mit dem das Label @home, ein Bienenkorb vor blauem Grund, verbunden bleiben soll.

=

Da die Gruppe vorwiegend aus Künstlerinnen und Künstlern besteht, die sich intensiv mit audiovisuellen Problemen befassen, stand zunächst die Frage zur Diskussion, ob der Projektraum als Kino eingerichtet werden sollte, in dem auf parallel sendenden Fernsehgeräten der Alltag der Gruppe nach Bern übertragen worden wäre. Eine zweite Idee für das Setting sah vor, den Projektraum in eine Werkstatt zu verwandeln, in der von den Künstlerinnen und Künstlern während der Ausstellung ein Magazin erarbeitet worden wäre, das die Geschichte des Kollektivs dokumentiert hätte. In beiden Fällen wäre auf die Präsentation von Artefakten verzichtet und auf den spezifischen Ortsbezug der Arbeit von @home hingewiesen worden. Doch schließlich konfrontierte mich die Gruppe, deren Arbeit ich als Kritik an der Ausstellungskunst wahrgenommen hatte, mit einem veritablen Ausstellungskonzept. Die fünf auf dem Boden verteilten blauen Kissen, die geöffneten Fenster, ein leichter Duft verdampften Honigs und die Reduktion der Beleuchtung auf zwei schwache Scheinwerfer führten dem Publikum am Tag der Eröffnung vor Augen, daß das Sichtbare in der Ausstellung Ergebnis einer Aktion war, von der es ausgeschlossen worden war. Zu sehen gab es in die Wände eingelassene Bergkristalle, ein Liniennetz in gelber Kreide, Bleistift- und Kohlezeichnungen. Das Setting gab vor, die Akteure hätten die Bühne eben verlassen. Was vorgegangen war, blieb für den Ausstellungsbesucher unnachvollziehbar. Gemeinsam ausgeführt wurden die in gelber Kreide nach langen Gesprächen entstandene Zeichnung und die danach in Bleistift auf die Wände übertragenen Diagramme von Bienentänzen, mit denen Kundschafter- den Sammelbienen besonders reichhaltige Nahrungsplätze anzeigen. Eine dritte, abschließend blind bei Nacht mit Kohle aufgetragene Zeichnungsschicht

war dem individuellen Ausdruck der an der Aktion beteiligten Gruppen-
mitgliedern vorbehalten. Die Gruppe, die bis dahin vor allem audiovisuell
gearbeitet hatte, hatte ihre eigenen künstlerischen Voraussetzungen über-
dacht und durch ihre Aktion versucht, Energie kollektiv und in möglichst
elementarer, experimenteller und direkter Form zu visualisieren.

=

Die Zusammenarbeit mit der Künstlergruppe @home hatte mit einem
Mißverständnis begonnen, das sich während der ganzen Vorberei-
tungszeit halten und sich erst kurz vor der Ausstellung allmählich klären
sollte. Während ich mich für den Lebenszusammenhang der Gruppe als
Arbeitszusammenhang interessierte und mich die Idee faszinierte, daß in
der Kunsthalle einmal keine Werke zu sehen sein würden, stattdessen aber
Informationen zu erhalten wären über die Lebenspraxis eines Kollektivs,
aus dessen Atelier bislang keine Artefakte im Umlauf gesetzt worden
waren, war @home immer davon ausgegangen, daß in Bern eine Ausstellung stattfinden werde. In der auf die Eröffnung folgenden Woche verteilten die Künstlerinnen und Künstler in ihrem Freundeskreis und an die
Ausstellungsbesucher Handzettel mit der Aufforderung, via Fax zu ihnen
zu sprechen oder die Ausstellung zu kommentieren. Die in der Kunsthalle
eingegangenen Nachrichten wurden täglich in der Ausstellung ausgehängt.
Am letzten Ausstellungstag fanden sich der Freundeskreis der Gruppe und
einige Ausstellungsbesucher in Bern ein, um sich die Ausstellung anzu-
sehen, die vielen ausgehängten Briefe zu lesen, um sich auf der Wiese hinter der Kunsthalle von @home bewirten zu lassen, am Synthesizer Geräu-
sche und Klänge zu erzeugen und auf den von der Gruppe mitgebrachten
Fellen in der Sonne zu liegen. Die Ausstellung war weniger als Installation
denn als Happening überzeugend, als Event, der die durch einen Ort und
eine Lebenspraxis außerhalb der Ausstellung bedingte Vernetzung der
Künstlerinnen und Künstler visualisierte.

Working in the Network

@ home

@home is the designation for a site, a homepage and the artists' collective that has occupied the place, working and playing with it in changing personnel constellations since 1995. The site is an old building with a long, colourful history, a structure that will probably be razed in the foreseeable future. Since renovated and modified to meet the needs of the artists' collective, the house and the buildings around it served before World War I as a transit station for emigrants from the Balkan states passing through Switzerland on their way to the United States. This functional role of the architectural complex is described in a novel written by a member of the family to which it belonged at the time. According to a spokesman for the group, the book was discovered in the abandoned building four years ago. The find was taken as an affirmation of the collective's choice of the space, as the book documents a pattern of functional continuity in the use of the house despite changing historical and social conditions. We find the following statement in a working paper issued by the group: "Recent years provide undeniable evidence of the changes in the aggregate state of our culture. This aspect is manifested in the increasing global nomadisation of culture and in the gradual emergence of an appreciation of the holotropic co-existence of interconnected realities and the demise of the concept of linear causality."

=

The history of @home dates back to the 1980s. The youth movement of the early 1980s, whose politically active participants occupied empty buildings, failed to find support for its objectives among the Swiss public. The following years witnessed the formation of collectives in many places, including a number of different locations in Basle, for which everyday life and especially the place in which one lived became the focal points of communal living. Political agitation gave way to a deliberately artistic and self-determined lifestyle inspired in an ideal sense by Joseph Beuys and his expansion of the concept of art into the idea of social sculpture. In its catalogue raisonné, @home lists the renovation and refurbishing of the house and the incorporation of furniture on wheels as the group's first work. Mobility, spontaneity, presence and amenability to change are among the key concepts in the philosophy of life and art practised by these young male and female artists. The @home studio has created practically no products but has realised numerous productions, including the Music Circus, a happening based on notes by John Cage that took place in the first year of the collective's existence. Situation-based environments are its most important form of expression. In 1996 and 1997, @home organised pop events under the banner of the Home Trainer Club, which were attended by various DJs, musicians and artists from Switzerland and abroad. In 1998, @home presented the Flugschule by transphere®, an installation comprising remote-controlled flying objects, light and slide projections on floating spheres, an interactive sound system and a Hare-Krishna dinner. The group organised a Mondmilchfahrt at Lac des Taillères, Switzerland in 1999. The event included a car trip, an outdoor happening in the snow and a karaoke party. @home unites Joseph Beuys's principle of freedom, defined by that artist in 1978 as a "concept of positive anarchy", with the spirit of the 1990s, which promoted a revival of 1970s fashion and design. @home displays aspects of personal life in public, exposes the collective in the individual and combines the personal sphere with the social. The staged character of the works produced by the collective is unmistakable. The artists are adamant in emphasising that @home represents a specific place with which the @home label – a beehive on a blue background – is inseparably associated.

=

Because the group is comprised primarily of artists intensely concerned with audio-visual issues, the first question that arose was whether the "Projektraum" should be set up as a cinema in which scenes from the everyday life of the group could be transmitted via parallel TV transmitters to Bern. A second proposal for the setting called for conversion of the "Projektraum" into a workshop in which the artists would prepare a magazine documenting the history of the collective for publication. Neither concept foresaw the presentation of artefacts, and each would have emphasised the specific site-orientation of the work of @home. Ultimately, however, the group, whose work I had seen as a form of criticism of exhibition art, came up with a genuine exhibition concept. The five blue pillows distributed over the floor, the open windows, a discreet scent of steamed honey and the reduction of illumination to two weak spotlights clearly indicated to the visitors who appeared on the opening day of the exhibition that the visible components of the exhibition were the results of an action from which they had been excluded. Viewers saw quartz crystals set in the walls, a network of lines in yellow chalk, pencil and charcoal drawings. The setting evoked the impression that the performers had just left the stage. Visitors had no way of determining what had gone on before. Working together, members of the group had executed the yellow-chalk drawing after a series of discussions and then the diagrams on the walls depicting bees' dances in which scout bees show collector bees the way to particularly abundant food supplies. A third layer of charcoal drawings done subsequently in the dark offered the group members participating in the action an opportunity for individual expression. The group, which had previously worked primarily with audio-visual resources, had re-evaluated its own artistic principles and attempted through its action to visualise energy collectively and in a form that was as fundamental, experimental and direct as possible.

=

Our work with the artists' group @home began with a misunderstanding that persisted throughout the preparation phase and was only gradually resolved shortly before the opening of the exhibition. While I had been interested in the group's living context as a working approach and was fascinated by the idea that, for once, instead of presenting works, the Kunsthalle would be making information available about a collective that had produced no artefacts for circulation, @home began its preparations on the assumption that an exhibition would take place in Bern. During the week following the opening, the artists distributed flyers to their friends and exhibition visitors, asking them to communicate with them by telefax or to comment on the exhibition. The messages received at the Kunsthalle were posted for view at the exhibition on a daily basis. On the last day of the show, some friends of the group and several visitors to the exhibition met in Bern to view the exhibition, read the posted letters, share food and drink served by @home on the lawn behind the Kunsthalle, makes sounds and noises on a synthesiser and lie in the sun on skins provided by members of the group. The exhibition was less convincing as an installation than as an event – a visual presentation of a network of artists linked by a particular place and a specific practical approach to life.

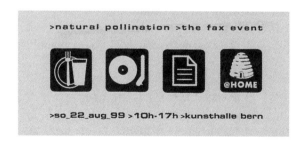

Weiß

Heinz Brand

Die Bestimmung des Ortes über eine Form von sichtbarer Abwesenheit – eines Menschen, einer Handlung oder eines Gegenstandes – zählt zu den Konstanten im Schaffen von Heinz Brand. 1977 verschickte der Künstler eine Karte mit der inzwischen legendären Erklärung, er habe einen autonomen Staat gegründet. Das Land umfaßte einen weiß gestrichenen, unmöblierten Raum im vierten Stock der Brunngasse 54 in Bern, so groß, daß der Freundeskreis des Künstlers gut darin Platz fand. Dieser Ort war betretbar, einige seiner Freunde waren sogar im Besitz eines eigenen Schlüssels zu diesem Land, doch auf den von Heinz Brand verschickten Drucksachen ist das Land stets als Leervolumen abgebildet, von außen photographiert durch das nach der Entfernung einer Zimmerwand sichtbar gewordene Balkenwerk. Mit der Wahl von Ornans, dem in der Franche-Comté gelegenen Geburtsort des französischen Malers Gustave Courbet (1819–1877), als Ort der vom Projektraum veranstalteten Ausstellung begab sich Brand auf besetztes Gelände. Das für seine Person und sein Schaffen charakteristische Wechselspiel von Forderung und Entzug hatte schon Courbet meisterhaft beherrscht. Als großer Maler und gerissener Provokateur des guten Geschmacks schrieb er sich unauslöschlich in das kulturelle Gedächtnis ein. Courbet ist noch heute in Ornans, wo er zeitlebens ein Atelier unterhalten hat, allgegenwärtig. Sein Geburtshaus, im Ortskern an der Loue gelegen, dient heute als Museum. Wer das über dem Hafen von La Tour-de-Peilz in der Waadt gelegene Schweizer Refugium seiner letzten Lebensjahre gesehen hat, das Courbet nach dem Scheitern der Pariser Kommune von 1871 auf der Flucht vor den französischen Behörden bezogen hatte, der kann aus der Ähnlichkeit der architektonischen Anlage und ihrer Dimensionen nur auf die große Bedeutung schließen, die das Geburtshaus in Ornans für den Künstler über all die Jahre behalten hatte. Eine Wanderung durch das Vallée de la Loue, sagt Heinz Brand, ist wie eine Bewegung durch eine Landschaft Courbets.

=

Zu den Arbeiten von Heinz Brand über den Ort und die Bestimmung des Ortes über eine Form von sichtbarer Abwesenheit – eines Menschen, einer Handlung oder eines Gegenstandes – zählen etwa der Abdruck eines nackten weiblichen Gesäßes auf einem Kegel weißen Marmorstaubs und das in Ornans als Edition entstandene *Unsichtbare Selbstbildnis* (1999): die Photographie eines leeren Steinsockels vor einem grünen Hintergrund, scharf fokussiert auf die kurz vor der Aufnahme vom Sockel gestiegene Figur. Die von Heinz Brand als *Luftlinien* bezeichneten Arbeiten bestehen aus den beiden Hälften eines aufgesägten Steins von besonderer Schönheit, die an unterschiedlichen Standorten so aufgestellt werden, daß die beiden Schnittflächen in einer imaginären Linie liegen. Imaginäre Linien dienen in diesem Schaffen der Ortsbestimmung. Was heißt das genau? Vor einigen Jahren zeigte Brand in einer Ausstellung ein Aquarium aus schwarz eingefärbtem Glas (*Blind cave fish,* 1994), in dem sich, unsichtbar für die Besucher, ein Fisch im Wasser bewegte. Brand bezeichnete die Gesamtheit der Bewegungen dieses Tieres als unsichtbare Zeichnung und sprach von einer Zeichnung in Schwarz. Es handelte sich um einen blinden Fisch, den Mexican Blind Cave Fish, dessen Augen sich im Laufe der Evolutionsgeschichte zurückgebildet haben, ein Fisch, der in Höhlen lebt und ohne Licht auskommt. Man konnte den im Aquarium schwimmenden Fisch durch das schwarz eingefärbte Glas nicht sehen, sich aber die Bewegungen des Fisches vorstellen.

=

Heinz Brand zeigte in Ornans mehrere Arbeiten. Die Installation im hellblau tapezierten Zimmer 33 des Hôtel de France war eine situationsspezifische, skulpturale Fassung der seit den siebziger Jahren entstehenden Fernseh-Installationen. Vor das in seinem Zimmer vorgefundene Fernsehgerät legte der Hotelgast Brand ein Stück einer dünnen, weißen Marmorplatte. Das Fernsehgerät sendet flüchtige Bilder, die durch den Marmor gefiltert, weich und unscharf erscheinen. Die Hinzufügung eines einzigen Elements, der Marmorplatte, verwandelte das Ensemble, bestehend aus Fernsehgerät und Hotelkühlschrank, in eine skulpturale Arbeit, die nicht nur die melancholische Eleganz barocker Skulpturen konnotierte, sondern durch die Funktion des Marmors als Bildschirm die Unbeständigkeit der gesendeten Fernsehbilder erst sichtbar machte. Für die im selben Hotelzimmer ausgestellten weißen Bilder bediente sich der Künstler zweier Reproduktionen nach Gemälden Courbets, einer Waldlandschaft mit Wasserfall und einem stürmischen Seestücks. Er kopierte die beiden Reproduktionen seitenverkehrt auf dünnes Fotopapier und ließ die Papiere umgekehrt auf Aluminiumtafeln aufziehen. Brand erzeugte Bilder, die sich auf der Rückseite des Bildträgers befinden und durch diesen hindurchscheinen. Die beiden Arbeiten wurden der Logik des befristeten Zimmerbezugs entsprechend nicht aufgehängt, sondern mobil präsentiert. In Umkehrung der von Courbet angewandten Technik, die weiße Farbe abschließend aufzutragen, bildet sie auf den Arbeiten von Brand die Grundfarbe. Die Absenz der Zeichnung ist die Bedingung ihrer Erscheinung. Weitere Arbeiten waren außerhalb des Hotels zu sehen. In einem schön gelegenen, verwilderten großen Garten über der Altstadt wurde mittels einer minimalen Veränderung durch den Künstler nicht nur ein Raum der Kunst ausgezeichnet, sondern auch Zeit als plastische Kraft sichtbar gemacht. Der von einer hohen Steinmauer umgebene Garten umfaßt einen verlassenen

Pavillon und einen Brunnen und war früher in verschiedene, durch die ursprüngliche Bepflanzung eindeutig gekennzeichnete Nutzungszonen gegliedert, die im gegenwärtigen Zustand des Gartens kaum mehr voneinander unterschieden werden können. In den von alten Obstbäumen durchsetzten Wiesen wachsen Brombeeren, die Randzonen verwalden, in den Kräuterbeeten blühen Blumen, und die Reben bleiben seit vielen Jahren ungeschnitten. Vor dem verfallenden Pavillon gibt es eine Pergola mit einem runden, von Moos überwachsenen Steintisch. Heinz Brand hat den Boden innerhalb der Pergola von Scherben, Laub, Steinen und dürrem Holz bis auf den ebenfalls moosbewachsenen Grund gereinigt und durch diesen einfachen Eingriff in die Ordnung des Gartens den Steintisch optisch untersockelt. Er löste ein Bild heraus, das Zeit als plastische Kraft evozierte und zugleich sichtbar machte, daß deren Wirkung einen Moment lang angehalten worden war. Dieser Ort ist einer der Bilderinnerung. Wind und Regen haben das Werk ausgelöscht. Der in diesem Garten in Ornans ausgezeichnete Raum war nicht nur eine orts-, sondern auch eine zeitspezifische Arbeit, denn strenggenommen konnte sie nur im Modus ihrer Veränderung wahrgenommen werden.

=

Mit der Verlegung der Ausstellung aus der Kunsthalle Bern nach Ornans wollte ich einem bekannten Künstler weniger eine Ausstellung als die Realisierung einer ortsspezifischen Arbeit oder Werkgruppe ermöglichen. Während der Dauer dieser Ausstellung in Ornans war in der Kunsthalle Bern die in Ornans entstandene Edition ausgestellt. *Das unsichtbare Selbstbildnis* (1999) sollte auf die außerhalb des Hauses stattfindende Veranstaltung hinweisen. Die Arbeiten von Heinz Brand basieren auf Ideen, doch im Unterschied zu anderen konzeptuellen Werken, die auf der begrifflichen Ebene wirksam werden, überzeugen seine Arbeiten durch ihre visuelle Präsenz. Mit der Ausstellung in Ornans war nicht weniger angezeigt als eine Ausstellung über die Poesie des Ortes.

White

Heinz Brand

The definition of a place in terms of a visible absence – of a person, an action or an object – is a constant in the art of Heinz Brand. In 1977, the artist sent out a card bearing the now legendary announcement that he had established an autonomous state. His country encompassed an unfurnished, white-painted room on the fifth floor of the building at Brunngasse 54 in Bern, a space large enough to accommodate the artist's friends. The site was open to access, and some of his friends even had keys to the country, yet the printed materials mailed out by Heinz Brand always depict it as an empty space, photographed from outside through the beams exposed after removal of one of the room walls. In selecting Ornans, the birthplace of the French painter Gustave Courbet (1819–1877) in the province of Franche-Comté, as the site of the exhibition organised for the "Projektraum", Brand entered occupied terrain. Courbet himself was a master of the interplay of demand and denial that is characteristic of Brand's own life and work. He made an indelible mark in the cultural memory as a great painter and a cunning provocateur of good taste. Courbet remains a ubiquitous presence in Ornans, where he maintained a studio throughout his lifetime. The house in which he was born, situated in the heart of town on the banks of the River Loue, is now a museum. For anyone who has seen the Swiss refuge above the harbour of La Tour-de-Peilz in French-speaking Switzerland where he spent he last years of his life, a domicile to which he retreated to escape the grasp of French authorities following the failure of the Paris Commune in 1871, the similarities of architecture and dimension between Courbet's last home and his birthplace in Ornans – and thus the meaning his birthplace held for him throughout his life – are impossible to overlook. A walk through the Vallée de la Loue, as Heinz Brand tells us, is like a stroll through a Courbet landscape.

=

Examples of works by Heinz Brand on the theme of place and its definition in terms of a form of visible absence – of a person, an action or an object – include the imprint of a woman's naked buttocks on a cone of white marble powder and the Invisible Self-Portrait *(1999) created as an edition in Ornans: the photograph of an empty stone pedestal in front of a green background, focused sharply on the figure that stepped down from the pedestal just before the exposure was made. The works Heinz Brand calls* Luftlinien *(Direct Lines) are composed of the two halves of particularly beautiful stones sawed apart, each half placed at a different location in such a way that the two cut surfaces lie along an imaginary line. In* these works, imaginary lines serve to define place. But what does that mean? Several years ago, Brand exhibited an aquarium made of black-tinted glass containing water and a fish which viewers could not see (Blind Cave Fish, *1994*). Brand referred to the totality of the movements of the fish as an invisible drawing, calling it a drawing in black. The fish was a Mexican blind cave fish, a species whose eyes have disappeared over the course of evolutionary history, an animal that lives in caves and requires no light. Although it was impossible to see the fish swimming in the aquarium through the black-tinted glass, one could easily imagine its movements.

=

Heinz Brand exhibited a number of works in Ornans. The installation in Room 33 of the Hôtel de France, a room with light-blue wallpaper, was a site-specific sculptural version of the TV installations he had been creating since the 1970s. Hotel guest Brand laid a thin, white marble slab in front of the television set in his room. The TV set transmitted fleeting images which, filtered through the marble, appeared soft and out of focus. The addition of a single element – the marble slab – transformed the entire ensemble consisting of a TV set and a hotel refrigerator into a work of sculpture that not only suggested the melancholy elegance of baroque sculptures but also, through the use of marble as a screen, exposed the inconsistency of the transmitted TV images to view. For the white paintings also exhibited in the hotel room, Heinz Brand used reproductions of two paintings by Courbet, a forest landscape with waterfall and a storm-lashed seascape. He copied the two reproductions, left-right reversed, onto thin photographic paper and had these printed in reverse on aluminium plates. Brand produced images on the back of his image-bearing media which shine through these surfaces. In keeping with the logic of the temporary occupation of the room, the two works were not hung on the walls but instead presented as mobile pieces. In a reversal of Courbet's technique of applying white paint in a final step, Brand used white as the ground in his works. The absence of the drawing is the factor that determines their manifest appearance. Other works were exhibited outside the hotel. In a beautifully situated, large, overgrown garden on the heights above the old city centre, the artist employed a minimal modification to identify a setting for art while rendering time visible as a sculptural force. Surrounded by a high stone wall, the garden encloses an abandoned pavilion and a fountain. It is divided into several different areas of use characterised by differences in the original planted vegetation but hardly distinguishable from one another given the present state of the garden. Blackberries grow in the meadows spotted with old fruit trees. Trees have begun to grow in the fringe areas, flowers bloom in the old herb beds and the vines have not been pruned for many years. Outside the derelict pavilion is a pergola with a round, moss-covered stone table. Heinz Brand swept the ground inside the *pergola free of shards, leaves, stones and withered branches and twigs, a simple intervention in the order of the garden through which he created a visual frame or base for the stone table. He isolated an image that evokes a sense of time as a sculptural energy, while showing that its effect had been suspended for a brief moment. This place is indeed one of visual memory. Wind and rain have washed the work away. The place identified in this garden in Ornans was not only a site-specific work but a time-specific one as well. Strictly speaking, it could only be appreciated within the context of its own change.*

=

The decision to transfer the exhibition from the Kunsthalle Bern to Ornans was prompted less by the idea of presenting an exhibition for an important artist than by the prospect of realising a site-specific work or group of works. The edition created in Ornans was exhibited at the Kunsthalle Bern during the Ornans show. The Invisible Self-Portrait *(1999) was meant to call attention to the event that took place outside the institution. Heinz Brand's works are based upon ideas, yet unlike other conceptual works whose effects unfold at the cerebral level, his art has a striking visual presence. The presentation in Ornans was nothing less than an exhibition about the poetry of place.*

Nahtstelle

Eran Schaerf

Der französische Schriftsteller Marcel Proust vergleicht die Arbeit des Schriftstellers mit der eines Schneiders, der ein Kleidungsstück aus bereits zugeschnittenen Teilen zusammennäht. Montage und Collage sind künstlerische Verfahren, die mit der Industrialisierung auftauchen, sich während der Moderne in der Bildenden Kunst durchsetzen und während des gesamten 20. Jahrhunderts Anwendung finden. Der französische Strukturalist Claude Lévi-Strauss hat in diesem Zusammenhang von einer doppelten Artikulation gesprochen. Um von einer doppelten Artikulation sprechen zu können, müssen die eingebauten Elemente in einem künstlerischen Werk als solche erkennbar bleiben. Claude Lévi-Strauss hat sie bei Poussin nachgewiesen, der nach antiken Vorbildern in Wachs modellierte Figuren als Modelle für seine Gemälde verwendete. Obschon das Verfahren der doppelten Artikulation heute in der Kunst eher die Regel denn die Ausnahme bildet, gibt es nur wenige Künstler wie Eran Schaerf, dessen Werk als Reflexion auf diesen Modus verstanden werden kann.

=

Eran Schaerf erarbeitete eine Ausstellung, die zwar thematisch an frühere Arbeiten anschließt, in der der Künstler aber durch das Hinzuziehen von Lichtbildern und Textfragmenten einen neuen Modus der Visualisierung erprobte. Am Eingang der Kunsthalle bekamen alle Besucherinnen und Besucher der Ausstellung eine Eintrittskarte, die mit jeweils verschiedenen Textfragmenten bedruckt war. Die meisten Besucherinnen und Besucher werden ihren Text nicht erst in der Installation, sondern noch an der Kasse oder auf dem Weg zur Ausstellung im Untergeschoß gelesen haben. Die aufgedruckten Texte sind Ausschnitte aus dem Skript, das die Konzeption der Installation im Projektraum bestimmt hat. Im Ausstellungsraum waren vier Tische aufgebaut, von denen aus Projektionen erfolgten. Jeder Tisch bestand aus einer quadratischen Holzplatte auf zwei Holzböcken. Aus den vier Tischplatten hatte Schaerf je eine Kreisscheibe herausgeschnitten, auf die Drehbühnen gelegt, die er zuvor ins Zentrum der Tische gerückt hatte, und die Projektoren darauf abgestellt. Die Kabel hatte er durch diese eine Öffnung im Tischblatt hindurch auf den Boden und weiter zu einem zentralen Stromanschluß geführt. Von den vier Tischen aus wurden nun, um jeweils eine Position in der Anordnung versetzt, dieselben Lichtbilder in derselben Reihenfolge auf die Wände des abgedunkelten Raumes übertragen. Eran Schaerf hatte schon den Besucherinnen und Besuchern früherer Ausstellungen mit dem Verkauf der Eintrittskarte auch eine Rolle verkauft; nun beabsichtigte er dem Publikum mit dem Eintritt einen Ausschnitt einer Geschichte zu verkaufen, die in der Installation im Projektraum mittels Lichtbildern in der einen oder anderen Form erzählt wurde. Die Drehrichtung der Projektoren war dieselbe, nicht aber die Drehgeschwindigkeit. Dieser minimale Unterschied erzeugte stets neue Bildkombinationen. Die Bilder blieben immer dieselben, doch Größe, Schärfe, Aufscheinort und Kombination mit anderen Bildern änderten sich und damit auch die Geschichte, selbst für zwei Besucher, die sich gleichzeitig in der Installation aufhielten. Die Nacherzählung derselben Bilder, vorbereitet durch zwei verschiedene Texte, erzeugt unterschiedliche Geschichten.

=

Das Verhältnis von Bild und Wort sowie die Frage der Ortsspezifik in der Wirklichkeitswahrnehmung und -darstellung finden sich im Schaffen von Eran Schaerf seit langem thematisiert. In seinen für den Projektraum konzipierten Arbeiten bezieht er sich auf einen französischen Salon des 18. Jahrhunderts, der als Geschenk der Familie Rothschild im Israel Museum in Jerusalem rekonstruiert wurde. Die Möbel, mit denen dieser Salon eingerichtet wurde, stammen zwar alle aus dem 18. Jahrhundert, bildeten zuvor aber keine Einrichtung. Sie wurden für diesen Zweck erst zusammengetragen und in das Museum versetzt. Die einzelnen Elemente sind authentisch, und dennoch ist das *Rothschild Zimmer* natürlich kein französischer Salon des 18. Jahrhunderts. Die Arbeiten von Schaerf handeln zwar von spezifischen Orten, wie es die ortsspezifische Kunst seit den sechziger Jahren tut, doch sie erzählen die Orte, das heißt, übersetzen sie in Sprache und Bild. Einer dieser Orte in unserer Arbeit ist das Israel Museum, dort das erwähnte *Rothschild Zimmer,* dort das einzelne Möbelstück und dessen Provenienz. Eine andere Schnittstelle ist Stanley Kubricks Film *2001: A Space Odyssey* von 1968, in dem der Mensch als Marionette der Technologie in einen kosmischen Zusammenhang gestellt wird. Der Bezugsort dort ist ebenfalls ein Raum im Stil des 18. Jahrhunderts, der von Kubrick ins Bild gerückt wurde. Auf einer der Eintrittskarten von Eran Schaerf heißt es denn auch, »einer Meldung der Nachrichtenagentur MGM zufolge« sei »der im Weltraum verlorengegangene Zukunftsforscher der Firma Kubrik zuletzt in einem französischen Salon des 18. Jahrhunderts gesehen« worden. Auf einem weiteren Text, der ebenfalls auf den Eintrittskarten abgedruckt ist, läßt der Künstler in Israel Baron Edmond de Rothschild das Zimmer im Stil des 18. Jahrhunderts einrichten, damit der verschollene Wissenschaftler wiederkehren kann. Eine dritte Schnittstelle bildet die »Samson-Truppe« der israelischen Armee. Diese Spezialeinheit operiert nicht in Uniform, sondern einsatzbezogen in entsprechender Verkleidung. Es gibt weitere Schnitt- und Nahtstellen in dieser Arbeit, deren Thematisierung erlauben würde, die Erzählung auszuweiten. Die Geschichte könnte ausführlicher, aber nicht vollständiger und keinesfalls

ganz erzählt werden, selbst dann nicht, wenn dafür alle auf den Eintrittskarten abgedruckten Texte und alle von den sich langsam drehenden Projektoren auf die Wände übertragenen Bilder hinzugezogen würden. Der
Grund ist der, daß die Geschichte weder als vollständiger Bildablauf konzipiert noch als abgeschlossener Text verfaßt worden ist.

=

Die Arbeit mit Bild- und Wortmodulen, deren Variation und Abwand-
lung hat im Werk von Eran Schaerf einen hohen Stellenwert. Eran
Schaerf entwirft installative Arbeiten für den Kunstkontext, arbeitet aber
auch in den Bereichen Buch, Film (zusammen mit der Schriftstellerin Eva
Meyer) und Hörspiel. Stets handeln seine Arbeiten von der feinen Nahtstelle zwischen »Wirklichkeit« und »Fiktion«, führen also immer auch über
ihre eigene Ordnung hinaus. Eine Zeitungsnotiz wie jene, die berichtet, daß
ein großes deutsches Unternehmen sich vertraglich verpflichtet hat, die
Rekonstruktion des legendären *Bernstein-Zimmers* im Katharinenpalast
bei St. Petersburg zu finanzieren, beschreibt eine solche Nahtstelle. Das
Bernstein-Zimmer stammt wie das *Rothschild Zimmer* aus dem 18. Jahrhundert. 1716 hatte der preußische König Friedrich Wilhelm I. das 1701
geschaffene und seit dem Zweiten Weltkrieg verschollene Kabinett dem
russischen Zaren Peter dem Großen geschenkt. Eine andere Nahtstelle ist
die eben erschienene Biographie des ehemaligen amerikanischen Präsidenten Ronald Reagan von Edmund Morris, die derzeit in den Vereinigten
Staaten für Debatten sorgt. Morris, der vor 14 Jahren, noch während der
Amtszeit von Ronald Reagan, an dessen Biographie zu schreiben begonnen
und nicht nur Zugang zu privaten Aufzeichnungen Reagans hatte, sondern
auch an Sitzungen des Kabinetts teilnehmen durfte, hat sich nicht auf
die Darstellung und Kommentierung dieser Quellen beschränkt, sondern
neben tatsächlichen Vorkommnissen aus dem Leben Ronald Reagans auch
erfundene Begegnungen und Ereignisse beschrieben und diese mit fiktiven
Quellen belegt, die im Anhang des Buches ununterscheidbar von den wirk-
lichen Quellen aufgeführt sind. Die Arbeiten von Eran Schaerf oder die
erwähnte Biographie lehren uns, selbst Tatsachen als Übersetzungen zu
behandeln, nicht nach dem Ursprung oder der Wahrheit, sondern nach
der Herkunft und der Funktion von Bildern und Worten zu fragen, selbst
oder gerade dann, wenn ihre Bedeutung gegeben erscheint. Nicht nur im
Museum, sondern auch anderswo, so jedenfalls verstehe ich in der Raumarbeit von Schaerf die Referenz auf die »Samson-Truppe«, jene Spezialeinheit der israelischen Armee, die nicht in Uniform operiert, sondern einsatzbezogen in entsprechender Verkleidung.

@ home
Natural Pollination, 1999

@ home

Natural Pollination > the fax event, 22. August 1999

Heinz Brand
Installation o.T., 1999

Heinz Brand
Ein Weltbild in ein anderes hineingestellt, 1999

Heinz Brand
Whites, 1999

Eran Schaerf
Scenario Data # 32, 1999

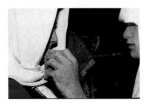

Eran Schaerf
Scenario Data # 32, 1999

Eran Schaerf
Scenario Data # 32, 1999

Anya Gallaccio
falling from grace, 2000

Anya Gallaccio
falling from grace, 2000

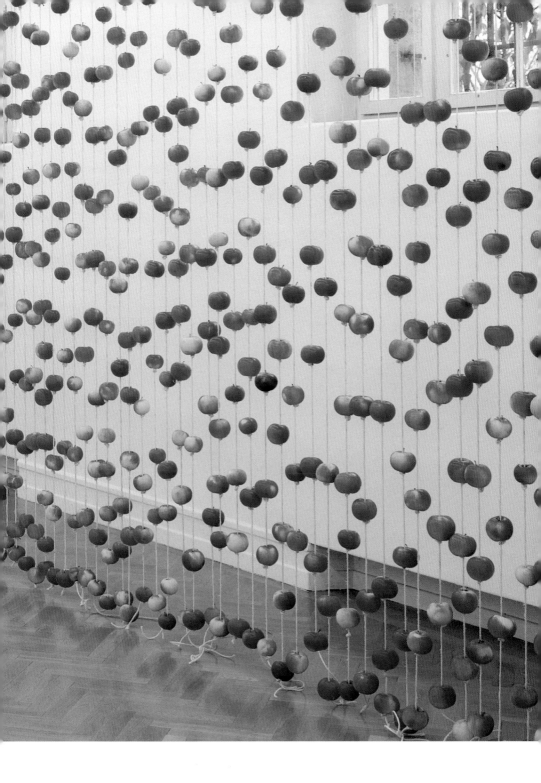

Anya Gallaccio
falling from grace, 2000

The Seam

Eran Schaerf

*French author Marcel Proust once compared the writer's craft to that
of a tailor, who makes articles of clothing by sewing together pieces
already cut to size and shape. The techniques of montage and collage first
appeared in art in the early years of industrialisation. They were firmly
established in visual art during the modern period and have been used by
artists throughout the 20th century. The French Structuralist Claude Lévi-
Strauss used the term "dual articulation" in this context. The concept of
dual articulation presupposes that the discreet elements incorporated
into a work of art remain recognisable as such. Claude Lévi-Strauss found
it in the work of Poussin, an artist who used wax figures based on ancient
sources as models for his paintings. Although the technique of dual articulation remains the rule rather than the exception in contemporary art,
there are but a few artists – among them Eran Schaerf – whose work can
be read as a reflection on this aesthetic mode.*

=

*The exhibition conceived by Eran Schaerf related thematically to
earlier works. Here, however, the artist explored a new mode of visualisation by integrating slides and text fragments into his work. Every visi-
tor to the exhibition received an admission ticket bearing one of several
different printed fragmentary texts at the entrance to the Kunsthalle. Most
visitors probably did not wait until they entered the installation to read the
text on their tickets but did so at the cashier's window or while on their way
to the exhibition on the basement floor. The printed texts were excerpts
from a script upon which the conception of the installation in the "Projektraum" was based. The projections were made from four tables constructed
of square wooden panels laid on sawhorses placed at different points in the
darkened room. Schaerf cut a circular disc from each of the four tables, laid
the discs on the revolving platforms he had previously positioned at the
centre of each tabletop and placed the projectors on them. He fed the cables
through the single opening in the tabletops and along the floor to a central
power outlet. The same slides were projected in exactly the same sequence
from each of the four tables onto the walls, the second, third and fourth projectors each beginning with a slide one step further along in the sequence.
In earlier exhibitions, Eran Schaerf had also sold visitors a role to play
along with their admission tickets. Now his idea was to sell the ticket-
purchasers a story narrated in one form or another in the slide show
presented at the exhibition. All of the projectors revolved in the same direc-
tion but at different speeds This small difference generated constantly*

Einer Meldung der Nachrichtenagentur M G M zufolge, wurde der im Weltraum verlorengegangene Zukunftsforscher der Firma Kubrik zuletzt in einem französischen Salon des 18. Jahrhunderts gesehen. Weltweit werden epochale Räume gesucht.

Die amerikanische Weltraum-Agentur Odyssey vermutet den verlorengegan-

Die amerikanische Weltraum-Agentur O d y s s e y vermutet den verlorengegangenen Zukunftsforscher der Firma Kubrik in einem französischen Salon des 18. Jahrhunderts. Sie ersuchte den israelischen Geheimdienst um Hilfe. Unterdessen erklärte der französische Baron Edmond de Rothschild seine Absicht, dem israelischen Museum für epochale Räume einen französischen Salon des 18. Jahrhunderts zu schenken, um das Ereignis der Wiederkehr des Zukunftsforschers in Israel stattfinden zu lassen.

In Paris gab der Sprecher der Familie Rothschild bekannt, daß der Baron

In Paris gab der Sprecher der Familie R o t h s c h i l d bekannt, daß der Baron mit der Einrichtung eines französischen Salons im israelischen Museum für epochale Räume keine exakte Rekonstruktion beabsichtigt, sondern ein Bühnendekor, in dem sich ein Mensch vor zweihundert Jahren zuhause gefühlt haben würde. Eine Inszenierung der Wiederkehr des verlorengegangenen Forschers für die Presse schloß der Sprecher nicht aus.

In einer Sondersitzung beschloß die weltweit vernetzte Napoleonische

In einer Sondersitzung beschloß die weltweit vernetzte Napoleonische Gesellschaft ihr Jahrestreffen von Waterloo nach Jerusalem zu verlegen. „Die Einweihung eines französischen Salons im Heiligen Land", sagte ihr Sprecher, „ist für uns Hobby-Historiker und War-Gamer eine einmalige Gelegenheit, einen napoleonischen Feldzug im Nahen Osten zu inszenieren". Das Jahrestreffen der N a p o l e o n i s c h e n G e s e l l s c h a f t wurde vor allem durch die detailgenauen Nachstellungen napoleonischer Schlachten bekannt.

Einer Gruppe von Frauen mit palästinensischem Akzent wurde der Einlaß

Eran Schaerf

Scenario Data # 18–21, 1999

changing combinations of images. Although the pictures themselves were always the same, their size, sharpness of focus, projection point and combination with other images changed, thus altering the story as well – even for two visitors inside the installation at the very same time. Visitors prepared with two different texts retold entirely different stories on the basis of the same sequence of pictures.

=

Eran Schaerf's art has been concerned for some time with the relationship between visual images and words and with the question of site-specificity in the perception and representation of reality. In his works conceived for the "Projektraum", he alludes to an 18th-century French salon donated by the Rothschild family to the Israel Museum and reconstructed in Jerusalem. All items of furniture in the salon are from the 18th century, although they were never combined to furnish a room before. The pieces were collected and shipped to the museum for that purpose. Although the individual items are authentic, the "Rothschild Room" is, of course, not a real French salon from the 18th century. Schaerf's works do indeed deal with specific places, as site-specific art has done ever since the 1960s, but they "narrate" these places – that is, they translate them into words and images. One of the sites in the work in question is the Israel Museum, within it the "Rothschild Room" and within the room the individual piece of furniture and its provenance. Another point of intersection is Stanley Kubrick's film 2001: A Space Odyssey *(1968), which presents an image of the human being in a cosmic context as a puppet of technology. The place of reference in the movie is also an 18th-century-style room used as a setting by Kubrick. Printed on one of Eran Schaerf's admission tickets are the words: "according to a report filed by the news agency MGM, the future researcher from the Kubrik company who was lost in space was last seen in an 18th-century French salon." In another text, also printed on admission tickets, the artist informs readers that Baron Edmond de Rothschild had the room furnished in the style of the 18th century in Israel so as to enable the lost scientist to return. A third point of intersection is represented by the Israeli Army's "Samson Group". This special unit does not wear uniforms but uses disguises appropriate for specific operations instead. There are other seams and points of intersection in this work, and the narrative could be expanded by emphasising them. The story could be told in greater detail, but it could not be told more completely or brought to a conclusion, even with the help of all of the texts printed on the admission tickets and all of the images cast on the walls by the slowly revolving projectors. The reason is that the story was neither conceived as a complete sequence of images nor written as a text with a beginning and an end.*

=

Modules of imagery and language in different variations and forms of adaptation play a very important role in Eran Schaerf's art. Eran Schaerf creates installations for the art context but is also involved with book, film (together with the writer Eva Meyer) and radio play production. All of his works deal with the fine seam along the border between truth and fiction, and thus they all point beyond their own internal ordering system. Such a seam is described, for example, in a newspaper article in which we read that a large German corporation has agreed to finance the reconstruction of the legendary "Bernsteinzimmer" (Amber Room) in the palace of Catherine the Great near St. Petersburg. Like the "Rothschild Room", the "Bernsteinzimmer" has its origins in the 18th century. King Frederick-William I of Prussia made a gift of the showpiece room, a work completed in 1701, to Russian Tsar Peter the Great in 1716. The "Bernsteinzimmer" has been missing since the Second World War. Another seam is the recently published biography of former US President Ronald Reagan by Richard Morris, a book that has sparked controversy in the United States. Morris, who began work on the biography in 1985, while Reagan was still in office, not only had access to Reagan's personal records but also took part in cabinet meetings. In addition to citing and commenting on these sources and relating true stories from Reagan's life, Morris also described fictitious encounters and events, documenting them with non-existent sources listed along with the authentic materials in the appendix and virtually impossible to distinguish from them. The lesson we learn from both Eran Schaerf's works and the Reagan biography is that even facts themselves must be treated as translations, that it is important to ask not only about the origin or truth of visual images and words but about their sources and functions as well, even – or especially – when their meaning appears self-evident. This applies not only in the museum setting but elsewhere as well. Such, at any rate, is my reading of Schaerf's reference to the Israeli Army's clandestine "Samson Group" in his spatial installation.

Jetzt

Anya Gallaccio

Anya Gallaccio bezog sich in ihren frühen Arbeiten auf künstlerische Konzepte der sechziger Jahre und realisierte eine Reihe von Werken, für die Struktur- und Ordnungsvorstellungen bestimmend waren. Eine ihrer ersten Arbeiten trägt den Titel *Waterloo* und entstand für die von ihrem Studienkollegen Damien Hirst zusammengestellte Ausstellung *Freeze,* die 1988 in einem der alten Lagerhäuser in den Surrey Docks in London zu sehen war. Anya Gallaccio goß eine dünne Bleischicht in der Form eines Rechteckes auf den Hallenboden. Rückblickend zeigt sich, daß diese Arbeit auf die genau zwei Jahrzehnte zuvor entstandenen *Splashing Pieces* von Richard Serra antwortet, von denen eines 1969 in der Kunsthalle Bern in der Ausstellung *When Attitudes Become Form* zu sehen war. Während Serra wenig flüssiges Blei schnell und kraftvoll gegen die Wand geworfen hatte, schmolz Anya Gallaccio große Mengen des Metalls und überzog den Boden in einem zeitintensiven und als solchen in der Arbeit abgebildeten Prozeß mit einer metallenen Haut.

=

In vielen seither entstandenen Werken verwendete Anya Gallaccio organische Materialien. Die für den Projektraum entwickelte Installation mit Äpfeln gehört in diesen Arbeitszusammenhang. In ihrem Schaffen mit organischen Materialien (Früchte, Schnittblumen, Schokolade, Körpersäfte, Wasser, Mineralien u.a.m.) ist der Prozeß der Selbstverwandlung des Werkes zentral. Mit einem zusätzlichen Glas werden beispielsweise Blumen hinter schon bestehende Fenster oder verglaste Türöffnungen gepreßt, oder sie werden zwischen zwei starke Gläser gelegt, die sowohl flach auf dem Boden liegend als auch aufrecht an einer Wand installiert ausgestellt sein können. Die Blumen verwelken, trocknen aus oder zersetzen sich. Die Arbeit *Recall* (1993) bestand aus kostbar gerahmten flachen Glaskästen, die Harn, Speichel, Blut und Sperma enthielten. Die Rahmen stammten aus einem Museumsdepot und gehörten zu historischen Männerbildnissen, die während der Ausstellung ungerahmt im Depot lagerten. Der Schokoladenraum *Couverture,* den die Künstlerin 1994 entwarf, befand sich im Kellergeschoß eines leerstehenden Wohnhauses in Basel. Anya Gallaccio bestrich das verputzte Bruchsteinmauerwerk eines Raumes bis auf etwa zwei Meter Höhe mit einer dickflüssigen, warmen Masse aus Schokolade und Kokosfett. Der Keller war nur über eine steile Holztreppe zu betreten und in unterschiedliche Räume aufgeteilt, die aber nicht durch Türen voneinander abgetrennt waren. Obwohl sich die Künstlerin entschieden hatte, in dem vom Eingang am entferntesten gelegenen kleinen

Eckraum zu arbeiten, schlug einem schon auf der Treppe der süßliche, aber nicht sofort zu identifizierende Schokoladenduft entgegen. Eine von der Holzdecke hängende Glühbirne erhellte den bemalten Raum, in dem eine einfache, schmale weiße Holzbank stand. Nach einigen Wochen verloren sich die skulpturalen Eigenschaften der Arbeit: Das satte, speckige Braun des Schokoladenanstrichs wurde gräulich, und während des feuchten Frühsommers bildete sich zunächst grauer, dann in kurzer Abfolge gelber, oranger und schwarzer Schimmel. Das Werk büßte diese malerische Farbigkeit nach einem weiteren Jahr wieder ein, weißer Schimmel bedeckte großflächig die Wände. 1996 schließlich, im Jahr, in dem Gebäude und Werk einem Neubau weichen mußten, war der Verputz aus Schokolade und Kokosfett durch Mottenfraß weich und brüchig geworden. Wer die Ausstellung mehrmals besuchte, traf jedesmal auf eine andere optische, olfaktorische und atmosphärische Situation und ein neues, überraschendes Bild desselben Ortes.

=

Die Instabilität von Form, Material, Farbe, Geruch und Bedeutung ist bezeichnend für die Arbeiten von Anya Gallaccio. Diese Thematisierung von Vergänglichkeit ist ein Schnittpunkt zahlreicher Werke britischer Kunst der neunziger Jahre: Die Plastik *Self* (1991) von Marc Quinn besteht aus gefrorenem Blut – des Künstlers eigenem Blut. Die Vitrine dient zugleich als Sockel und Kühlfach. Während fünf Monaten in kleinen Mengen entnommen und in eine Gußform seines Kopfes gegossen, entspricht das entnommene Blutvolumen ungefähr jener Menge, die in einem menschlichen Körper zirkuliert. Zu den umstrittenen frühen Arbeiten von Damien Hirst zählen *The Physical Impossibility of Death in the Mind of Someone Living* (1991), ein in Formalin konservierter Tigerhai, sowie *A Thousand Years* (1990), eine Doppelvitrine, in der Fliegen aufwachsen, sich vermehren und verenden. Zwar visualisieren viele Arbeiten von Anya Gallaccio ebenfalls Gefährdung und Instabilität, doch meint Vergänglichkeit in ihrem Werk nicht Vergeblichkeit, sondern wird als Hinweis auf ein zyklisches Geschehen wahrnehmbar.

=

Die Künstlerin reagiert mit ihren Installationen einerseits auf die geografische, architektonische und institutionelle Situation und arbeitet andererseits mit dem kollektiven kulturellen Gedächtnis des Publikums, insbesondere mit kunsthistorischen Reminiszenzen und jenen Teilen des humanistischen Erbes, die nur noch schemenhaft in unserer Kultur weiterleben. Die Werke von Anya Gallaccio sind in einem situativen und kulturellen Sinne ortsspezifisch: Ich erinnere an eine Arbeit von 1997, die Anya Gallaccio *keep off the grass* nannte. Sie säte Blumen- und Gemüsesamen in Grasnarben, die von Skulpturen herrührten, welche einst im Rasen vor der Serpentine Gallery in London ausgestellt waren. Die sprießenden Blumen und Gemüse zeichneten nicht nur ein Ornament in den Rasen vor dem Museum, sondern gliederten den Außenraum und evozierten Erinnerungsbilder an zurückliegende Ausstellungen.

=

Im Projektraum ordnete Anya Gallaccio Äpfel zu einem Tableau. In Kisten angelieferte und der Künstlerin vom Verkäufer als Masse verrechnete Früchte wurden als Fläche wahrnehmbar, die wie ein transparenter Vorhang vor der Fensterfront des Seitenlichtsaals hing. Mit ihrer Arbeit *falling from grace* unterteilte Anya Gallaccio den Saal in einen langen und schmalen Korridor mit Tageslicht und einen zweiten, künstlich beleuchteten Raum. Die Äpfel waren in unregelmäßigen Abständen wie Perlen auf Schnüre aufgezogen. Das Gewicht der auf Doppelknoten sitzenden Früchte hielt die in einer Linie von der Saaldecke hängenden Schnüre straff. Die Schwerkraft war spürbar, erzeugt wurde der Eindruck fallender Äpfel. Dieses Bild war besonders stark, wenn man die Äpfel zusammen mit den winterlich kahlen Bäumen wahrnahm, die sich vor den Fenstern als Silhouetten abzeichneten.

=

Der Apfel ist ein Produkt, das in unserer Wirtschaft für den Massenkonsum angebaut wird, und hat gleichzeitig als Zeichen eminente kulturgeschichtliche Bedeutung. Der Apfelbaum ist eine der ältesten Kulturpflanzen und die Frucht in der abendländischen Kultur Sinnbild für die Erbsünde und deren Überwindung, für Macht, Optimismus, Perfektion und Heimat. Die kulturgeschichtliche Bedeutung eines Materials spielt für Anya Gallaccio in der konzeptionellen Phase des bildnerischen Prozesses eine wichtige Rolle, die Visualisierung einer Idee aber wird durch das optische Potential des Materials und die räumlichen Bedingungen bestimmt. Anya Gallaccio verarbeitete Äpfel ausschließlich einer Sorte, doch war jede einzelne Frucht in Farbe, Zeichnung und Umriß verschieden. An sonnigen Wintertagen wirkte die Arbeit wie ein von leichter Hand gemaltes Aquarell. Man kann *falling from grace* aber auch als Instrument zur Schärfung unserer Wahrnehmung verstehen: Die Inszenierung der Äpfel als Fläche im Raum war eine Aufforderung zum bewußten Sehen von Farben, Volumen, Glanzlichtern, Konturen, Schatten, Rhythmen und Wiederholungen sowie deren Veränderung in der Zeit. Dieser Prozeß ist kein linearer Ablauf, der mit Ausstellungsende abbricht. Die Künstlerin wird die Kerne nach Abbau der Installation aus den Früchten herauslösen und einpflanzen. Da der kultivierte Apfel sich generativ als Wildling vermehrt, wird aus *falling from grace* ein Garten wilder Apfelbäume von großer Sortenvielfalt heranwachsen. Arbeiten wie der entstehende Hain gestalten einen Ort und erhalten diesen lebendig, sind aber optisch als Kunstwerke kaum zu erkennen, sondern nur mehr konzeptionell zu beschreiben.

Now

Anya Gallaccio

Referring in her early works to artistic ideas of the 1960s, Anya Gal-
laccio realised a number of pieces in which concepts of structure and
order played a determining role. Waterloo, *one of her first works, was done*
for the exhibition Freeze *organised by her fellow student Damien Hirst in*
an old warehouse in London's Surrey Docks. Anya Gallaccio cast a thin
layer of lead into a rectangular form on the floor. In retrospect, we see in
this work an echo of the Splashing Pieces *created by Richard Serra pre-*
cisely two decades earlier, one of which was exhibited at the exhibition
When Attitudes Become Form *at the Kunsthalle Bern in 1969. In contrast*
to Serra, who had thrown a small amount of liquid lead in a quick, vigor-
ous motion against the wall, Anya Gallaccio melted down a substantial
quantity of the metal and covered the floor with a leaden skin in a time-
consuming process that was documented as such by the work itself.

=

Anya Gallaccio has used organic materials in many of her later works.
The installation with apples developed for the "Projektraum" belongs
to this context. In her work with organic materials (fruits, cut flowers,
chocolate, bodily fluids, water, minerals, etc.), the process of self-trans-
formation of the work is of central importance. Flowers, for example, were
pressed behind existing windows or glassed-in doorways with the aid of a
second pane of glass or laid between two thick panes of glass installed
either flat on the floor or upright against a wall for exhibition. The flowers
wilted, dried out or decomposed. The work entitled Recall *(1993) comprised*
glass cases, with precious old frames, containing urine, saliva, blood and
sperm. The frames came from the museum storage room and were bor-
rowed from historical portraits of men, which remained in storage without
frames during the exhibition. The chocolate room Couverture *created by*
the artist in 1994 was located in an empty apartment house in Basle. Anya
Gallaccio covered the stone walls of a room with a thick, warm paste con-
sisting of chocolate and coconut oil to a height of two metres from the floor.
The cellar, accessible only down a steep wooden stairway, was divided into
several different rooms, which were not separated from one another by
doors, however. Although the artist chose a small corner room located
farthest from the entrance for her work, visitors encountered the sweet but
not immediately identifiable smell of chocolate as soon as they began to
descend the stairs. A light-bulb suspended from the wooden ceiling illu-
minated the painted room, in which a simple, narrow wooden bench had
been placed. After several weeks, the sculptural qualities of the work had

disappeared. *The thick, greasy brown of the chocolate coating turned grey-ish, and mould began to form during the humid early summer months in a rapid succession of yellow, orange and black tones. This painterly colora-tion also faded away over the course of another year, as white mould spread over substantial portions of the walls. By 1996, the year in which the building and the work were to be demolished to make way for a new structure, the coat of chocolate and coconut oil had become moth-eaten, soft and brittle. Visitors who returned several times to the exhibition en-countered an altered visual, olfactory and atmospheric situation each time: a new and surprising image of one and the same place.*

=

The instability of form, material, colour, odour and meaning is a char-acteristic feature of the work of Anya Gallaccio. This focus on tran-sience is a point of intersection for numerous works of British art that originated during the 1990s: Marc Quinn's sculpture Self *(1991) consists of frozen blood – the artist's own blood. The showcase serves both as a ped-estal and a refrigeration unit. Drawn over a period of five months in small increments and poured into a mould taken from his head, the volume of blood collected corresponds roughly to the amount that circulates through the human body. Among the more controversial early works of Damien Hirst are* The Physical Impossibility of Death in the Mind of Someone Living *(1991), a tiger shark preserved in formaldehyde, and* A Thousand Years *(1990), a double-glass case in which flies grow, multiply and die. While many of Anya Gallaccio's works also visualise vulnerability and instabil-ity, transience does not signify futility in her art but is instead rendered visible as a sign of cyclical processes.*

=

On the one hand, the artist responds in her installations to the geo-graphic, architectural and institutional situations she encounters; on the other, she works with the collective memory of the viewing public and, in particular, with art-historical reminiscences and with those parts of the legacy of humanism that live on as mere shadows in our culture. The works of Anya Gallaccio are site-specific in a situative, cultural sense. I recall a work of 1997 entitled keep off the grass *by the artist. She sowed flower and vegetable seeds in bare spots in the grass left behind by sculptures once exhibited on the lawn of the Serpentine Gallery in London. The flowers and vegetables which sprouted there not only drew an ornamental configura-tion in the lawn outside the museum but also gave structure to the outdoor space and evoked memories of past exhibitions.*

=

In the "Projektraum", Anya Gallaccio arranged apples into a tableau. Delivered in boxes and sold to the artist as bulk by the grocer, the fruits were presented to viewers as a flat surface suspended like a trans-

parent curtain in front of the windows of the Seitenlichtsaal. In this work, entitled falling from grace, *Anya Gallaccio divided the room into a long, narrow corridor illuminated by daylight and a second, artificially lighted room. The apples were arranged in irregular intervals like pearls on strings. The weight of the fruits, held on the strings by square knots, kept the strings suspended in straight lines from the ceiling taught. The effect of gravity was readily apparent, and the image evoked was that of falling apples. This image became especially vivid when one saw the apples against the background of the leafless, winter-stripped trees that appeared as silhouettes outside the windows.*

=

The apple is grown for mass consumption in our economy, yet it has retained its eminent significance as a historical cultural symbol. The apple tree is one of the oldest of all cultivated plants, and, in Western culture, its fruit symbolises Original Sin and redemption, power, optimism, perfection and the concept of home. While the significance of a material in terms of cultural history plays an important part in the conceptual phase of Anya Gallaccio's creative work, the visual realisation of an idea is determined by the visual potential inherent in a given material and by the spatial conditions of a particular setting. Anya Gallaccio used only one type of apple, yet each fruit differed from the others in colour, surface marking and shape. On sunny winter days, the work had the look of a water-colour painted with a light hand. However, it is also possible to interpret falling from grace *as an instrument meant to sharpen our perceptive powers. The presentation of apples as a flat surface in space represents an appeal to viewers to open their eyes to colours, volumes, brilliant flashes of light, contours, shadows, rhythms and recurring phenomena as well as to changes over time. This process is not a linear sequence that comes to an end with the closing of the exhibition. When the installation is dismantled, the artist will remove the seeds from the apples and plant them. Because the cultivated apple multiplies as a wild plant as well,* falling from grace *will produce a garden of wild apple trees characterised by a broad diversity of species. Works like this orchard give shape and design to place, imbuing it with lasting life. In a visual sense, however, they are scarcely recognisable as works of art and can only be described in conceptual terms.*

Die Ausstellungen / The Exhibitions

Renate Buser, Marie Sester, Elizabeth Wright
21. Mai – 28. Juni 1998

Renate Buser

Geboren 1961, lebt und arbeitet in Basel

1997	The Banff Centre for the Arts, Banff/Alberta, Canada
1996–1997	Montreal
1996	Eidgenössischer Preis für Gestaltung (Experimentelle Photographie)
1992	Binz 39, Scuol GR
1990/94/97	Künstlerstipendium der Stadt Basel
1990	Atelierstipendium an der Cité Internationale des Arts, Paris
1985–1986	Accademia di belle Arti (Klasse von E. Vedova), Venezia
1982–1988	Schule für Gestaltung Basel

EINZELAUSSTELLUNGEN (AUSWAHL)

1999	Allee der Kosmonauten, Kunstraum Aarau (Katalog)
1998	Objects in the mirror may be closer than they appear, Southern Alberta Art Gallery, Lethbridge (Katalog)
1993	Funghi, Projektraum Zürich

GRUPPENAUSSTELLUNGEN (AUSWAHL)

1999	Der plastische Blick, Kunsthaus Langenthal
1999	Sichtweisen, Kunsthalle Palazzo, Liestal (Katalog)
1998	Renate Buser, Marie Sester, Elizabeth Wright, Kunsthalle Bern, Projektraum (Katalog)
1998	nonlieux, Kaskadenkondensator, Basel
1998	PleinAir Multimedia, Kraków
1997	Mois de la Photo, Galerie Occurrence, Montreal
1997	Objectif Lune, CAN, Neuchâtel
1997	Galerie M. Haldemann, Bern
1995	Blinzeln, Toit du Monde, Vevey
1993	überKreuz, Galerie Pankow, Berlin (Katalog)
1992	Galerie Luciano Fasciati, Chur

AUSGESTELLTE WERKE

— *Hotel Marriott, Hotel Crescent, La Canadienne, Westmount Square,* alle in Montreal, aus der Werkgruppe *Objects in the mirror may be closer than they appear,* 1997, Photographie auf Barytpapier

Marie Sester

Geboren 1955, lebt und arbeitet in Paris

2000	Artist in Residence, Künstlerhaus Dortmund
1999	Artist in Residence, Portland Institute for Contemporary Art, Portland
1998	Studienaufenthalt in New York
1996	Lauréate de la Villa Médicis Hors les Murs
1993	Lauréate de la Villa Kujoyama, Kyoto

EINZELAUSSTELLUNGEN (AUSWAHL)

2000	Portland Institute for Contemporary Art, Portland

2000	Musée Château, Musée de la Ville d'Annecy
1996	Centre d'Art Contemporain de Vassivière
1996	Château du Grand Jardin de Joinville
1995	Centre d'Art Contemporain de Rueil-Malmaison, Paris
1995	Galerie Psyché, Martigues
1992	Galerie Jade, Colmar
1991	Merle-Portalès et Associés, Paris
1990	Galerie Arthème, Palma de Mallorca
1990	Cloître des Dominicains, Guebwiller
1989	Galerie Arthème, Palma de Mallorca
1989	Merle-Portalès et Associés, Paris
1988	Galerie Luc Queyrel, Paris

GRUPPENAUSSTELLUNGEN (AUSWAHL)

2000	Künstlerhaus Dortmund (mit Thierry Fournier)
2000	Centre d'Art Contemporain de Pougues-les-Eaux
1999	New Langton Arts, San Francisco
1999	Souvenir Utopie: Architektur in der zeitgenössischen französischen Kunst, Stadthaus Ulm (Katalog)
1998	Renate Buser, Marie Sester, Elizabeth Wright, Kunsthalle Bern (Katalog)
1998	Artists Space, New York
1997	Kwangju Biennale, Kwangju (Katalog)
1996	Arcos da Lapa, Rio de Janeiro
1994	Hillside Gallery, Tokyo
1992	Merle-Portalès et Associés, Uzès
1990	Façades imaginaires, Laboratoire, Grenoble

AUSGESTELLTE WERKE

— *Appartement,* 1996, Plexiglas, 600 x 400 x 1,05 cm

Elizabeth Wright

Geboren 1964, lebt und arbeitet in London

| 1987–1990 | Royal College of Art, London |
| 1984–1987 | Birmingham Polytechnic, Birmingham |

EINZELAUSSTELLUNGEN (AUSWAHL)

1999	C579DJD, J839TVC, A896TL, Delfina, London
1996	Showroom, London (Katalog)
1995	Karsten Schubert, London
1994	Small appetite, Modern Art, London

GRUPPENAUSSTELLUNGEN (AUSWAHL)

1999	Wild Life, The Austrian Cultural Institute, London
1999	Ninenineninetynine, Anthony Wilkinson Gallery, London
1999	Shopping, FAT commission, Ganton Street, London
1999	Riverside 99, Norwich
1999	Appliance of Science, Frith Street Gallery, London
1999	Holding Court, Entwistle Gallery, London
1998	Fun de Siecle, Walsall Museum and Art Gallery, Walsall (Katalog)
1998	Resolute, Platform, London
1998	Thinking Aloud, Arts Council touring exhibition, Kettles Yard, Cambridge / Corner House, Manchester / Camden Arts Center, London
1998	The Vauxhall Gardens, Norwich Gallery, Norwich

1998	It took ages, Bricks and kicks, Wien
1998	Show us the money 2 and 3, Dukes Mews, London
1998	Renate Buser, Marie Sester, Elizabeth Wright, Kunsthalle Bern, Projektraum (Katalog)
1998	Real Life, Galleria Sales, Roma
1998	Host, Tramway, Glasgow
1998	Trash, Greene Naftali Inc, New York
1997	At one remove, Henry Moore Institute, Leeds (Katalog)
1997	With in these Walls, Kettles Yard, Cambridge (Katalog)
1997	A Sense of Scale, Ikon touring exhibition, Birmingham
1997	Thoughts, City Racing Gallery, London
1997	Belladonna, Institute of Contemporary Art, London
1997	Drei Zimmer für Julie Bondeli, Galerie Erika und Otto Friedrich, Bern (Katalog)
1996	British Waves, Roma
1996	A Glass of Water, 526 West 26th Street, New York
1996	Dinner, Cubitt Gallery, London
1996	Life is Elsewhere, Spacex Gallery, Exeter (Katalog)
1996	Ace, Arts Council touring exhibition, Hayward Gallery, London u. a.
1995	General release, La Biennale di Venezia (Katalog)
1995	Stoppage, C.C.C., Tours
1995	Life of it's own, British Council Window Gallery, Praha
1995	Imprint 1993, City Racing, London
1995	UK Wit and Excess, Contemporary Art Centre, Adelaide (Katalog)
1995	Mexico or Bust, Abercorn Place, London
1994	Dirty Holes, Modern Art, London
1994	Where you were even now, Filiale Basel (Katalog)
1994	Candyman II, London
1994	Complete, Filiale Basel
1994	Nichts als Inhalt, Museum für Gestaltung, Basel (Katalog)
1994	In Transit, Central Bus Station, Tel Aviv
1994	Television, Cubitt Gallery, London
1992	The Infanta of Castile, London
1992	Hit and Run, Tufton Street, London

AUSGESTELLTE WERKE
— *Stolen Bicycle,* enlarged by 165%, 1998, Aluminium, Stahl, Latex, Farbe

Katharina Grosse
4. Juli – 21. August 1998

Katharina Grosse

	Geboren 1961, lebt und arbeitet in Düsseldorf
1999–2000	Gastprofessur an der Hochschule für Künste Bremen
1999	Artist in Residence Chinati Foundation, Marfa, Texas
1998	Arbeitsstipendium Edenkoben des Kultusministeriums Rheinland-Pfalz
1997–1998	Gastprofessur an der Hochschule der Bildenden Künste Karlsruhe
1995	Arbeitsstipendium Kunstfonds Bonn e.V.
1993	Karl Schmidt-Rottluff Stipendium

1992   Villa Romana-Preis, Firenze
1982–1990   Kunstakademie Düsseldorf

EINZELAUSSTELLUNGEN (AUSWAHL)

1999   Kunstverein Bochum
1999   Galerie Barbara Gross, München
1999   Rheinisches Landesmuseum, Bonn (Katalog)
1999   Galerie Conrads, Düsseldorf
1998   Sightspacific, Kunstverein Bremerhaven
1998   Kunsthalle Bern, Projektraum (Katalog)
1998   Galerie Sfeir-Semler, Hamburg
1998   Todd Gallery, London
1998   Kunstverein Heilbronn
1997   Galerie Conrads, Düsseldorf
1997   Galerie Mark Müller, Zürich
1997   Petra Bungert Gallery, New York
1996   Galerie Sfeir-Semler, Kiel
1996   Overbeck-Gesellschaft, Lübeck (Katalog)
1995   Galerie Mark Müller, Zürich
1995   Galerie Conrads, Düsseldorf
1994   Altes Kunstmuseum, Bonn
1993   Galerie Sfeir-Semler, Kiel

GRUPPENAUSSTELLUNGEN (AUSWAHL)

1999   Zeitwenden, Kunstmuseum Bonn (Katalog)
1999   The Drawing Center, New York
1999   Ausloten: Fünf Positionen autonomer Malerei, Kunstverein Göttingen (Katalog)
1999   Viereck und Kosmos, Amden
1999   Chroma, Malerei der neunziger Jahre, Kunsthalle Nürnberg (Katalog)
1999   Auf die Wand, Galerie Mark Müller, Zürich
1998   9 + 1, Petra Bungert Projects, Bruxelles
1998   Every Day, 11th Biennale of Sydney (Katalog)
1998   Farbe, Galerie Monika Reitz, Frankfurt
1998   Sensation Farbe, Galerie Mark Müller, Zürich
1997   Topping Out, Städtische Galerie Nordhorn (Katalog)
1997   Augenzeugen, Kunstmuseum Düsseldorf (Katalog)
1997   Abstraction / Abstractions – Géométries Provisoires, Musée d'Art Moderne Saint-Etienne (Katalog)
1996   Farbe – Malerei der 90er Jahre, Kunstmuseum Bonn (Katalog)
1995   Karl Schmidt-Rottluff Stipendiaten, Kunsthalle Düsseldorf (Katalog)
1995   Karo Dame, Aargauer Kunsthaus Aarau (Katalog)
1994   Farbe benützen, Galerie Mark Müller, Zürich
1993   Junger Westen, Kunsthalle Recklinghausen (Katalog)

AUSGESTELLTE WERKE

— *Inversion*, 1998, Acryl auf Wand, 300 x 850 cm

Susanne Fankhauser
5. September – 18. Oktober 1998

Susanne Fankhauser

Geboren 1963, lebt und arbeitet in Basel

1999	Eidgenössischer Preis für freie Kunst
1990–1991	Atelierstipendium an der Cité Internationale des Arts, Paris
1989/93/97	Künstlerstipendium der Stadt Basel
1985–1989	Fachklasse für freies, räumliches Gestalten, Schule für Gestaltung, Basel

EINZELAUSSTELLUNGEN

1998	Kunsthalle Bern, Projektraum (Katalog)
1997	Räume für Kunst, Dorothea Deimann, Müllheim
1996	Kaskadenkondensator, Basel
1991	Kunstraum KIFF, Aarau

GRUPPENAUSSTELLUNGEN (AUSWAHL)

1998	Skulpturen und Neue Medien, Reinach (mit Sonja Feldmeier)
1994	Galerie Werkstatt, Reinach
1994–1996	Wanderausstellung Papierobjekte im Baltikum und in Rußland
1993	Mutationen, Filiale Basel
1992	Galerie Anton Meier, Genève

AUSGESTELLTE WERKE

— *Das Museum der Tiere,* 1998 (Bildlegenden auf den vorgefundenen Reproduktionen: Braco Dimitrijevic, *In Addition to Ruskin's Theory of Art,* 1989 / Mike Kelley, *Brown Star,* 1991 / Asta Gröting, *Paarverhalten,* 1989 / Lothar Baumgarten, *Entenschlaf (Wegwurf),* 1991–92 / Alex Flemming, *O.T.,* 1994 / Claudia Di Gallo, *Konkurs hippique,* 1996 / Jochen Gerz, *Die Schwierigkeit des Zentaurs beim vom Pferd steigen,* 1997 / Maurizio Cattelan, *Love Doesn't Last Forever* / Simon Beer, *Noli me tangere,* 1994 / Bertram Jesdinsky, *Giraffe,* 1989 / Robert Rauschenberg, *Odalisque,* 1955–58 / Susan Rothenberg, *Bärenhaut Teppich,* 1995 / Katharina Fritsch, *Pudel,* 1995 / Georg Ettl, *Pudel,* 1976 / Peter Fischli, David Weiss, Ausschnitt aus der Installation im Schweizer Pavillon, Biennale Venedig, 1995 / Marie José Burki, *Les chiens,* 1994 / Jeff Koons, *Poodle,* 1991 / Wim Delvoye, *Pigs, Life Tattooed Pigs,* 1996 / Charles Ray, *Revolution Counter Revolution,* 1990 / Wastijn & Deschuymer, *»R.O.Y.G.B.I.V. II«* / Stephan Balkenhol, *Overleaf,* 1991 / Katharina Fritsch, *Elefant,* 1987 / Heidemarie von Wedel, *O.T. (Elch),* 1994 / Paul Thek, *Chair,* 1968 / Andreas von Weizsäcker, *Objekt o.T.* / Maurizio Cattelan, Enzo Cucchi, Ettore Spaletti, Gemeinschaftsarbeit im italienischen Pavillon, Biennale Venedig, 1997 / Bruce Nauman / Louise Bourgeois, *Nest* / Jean Luc Vilmouth / Wolfgang Müller, *Riesenbrillenalk,* 1990 / Niki de Saint-Phalle, *Pouf Serie B,* 1991 / Joseph Beuys, *Aktion Coyote,* 1974 / Abigail Lane, *Bloody Wallpaper and Stone Dog,* 1995 / Inspektion medizinische Hermeneutik, *She is young discourse,* 1990 / Ashley Bickerton, *Solomon Island Shark,* 1993 / Diego Giacometti, *Die Katze oder Maître d'hôtel,* 1969), Jumbo-Ink-Jet-Druck, vierfarbig, PVC-Blache, 300 x 965 cm

Sarah Rossiter
31. Oktober – 6. Dezember 1998

Sarah Rossiter

Geboren 1970, lebt und arbeitet in Brooklyn, New York
1997–1998   Parsons School of Design, New York
1993   Cooper Union School of Art, New York

EINZELAUSSTELLUNGEN (AUSWAHL)
1999   Yuri-G, Thomas Erben Gallery, New York
1998   Kunsthalle Bern, Projektraum (Katalog)
1998   Velocity, Thomas Erben Gallery, New York
1997   I Object, Thomas Erben Gallery, New York
1996/1997   In Amerika, Thomas Erben Gallery, New York
1993   A Bunch of Dogs, Cooper Union, New York

GRUPPENAUSSTELLUNGEN (AUSWAHL)
1999   Wall to Walrus: Furnishing and Related Installation, Inshallah, Los Angeles
1994   Women at Work, Roger Smith Gallery, New York
1993–1994   Word of Mouth, Royal Danish Academy, Copenhagen / Palais des Beaux
    Arts, Bruxelles / Friesenwall 120, Köln
1993   Andrea Rosen Gallery, New York
1992   Proposals for the Museum of Natural History, Upstairs at Pat Hearn
    Gallery, New York (organized by Mark Dion)
1992   Tattoo Collection, Andrea Rosen Gallery, New York
1992   Personals, Virtual Gallery at Dooley LeCappellaine, New York

AUSGESTELLTE WERKE
— *Velocity,* 1998: *Velocity* (Passage 1), *Velocity* (Passage 2), *Velocity* (Grid with Sais),
Velocity (Blue Stripe 1), *Velocity* (Grid with Sticks), *Velocity* (Orange Stripe), *Velocity*
(Drips with Nunchucks), *Velocity* (Yellow and Red Drips), *Velocity* (Brown Drips), C-Print,
Auflage: 6, je 102 x 76 cm
— Sitzbank: *250 Met* (Design Piero Lissoni und S. Sook Kim), Courtesy: Teo Jakob AG, Bern

Lee Bul
6. Februar – 28. März 1999

Lee Bul

Geboren 1964, lebt und arbeitet in Seoul
1982–1987   BFA in Skulptur, Hong-Ik University, Seoul

EINZELAUSSTELLUNGEN
2000   Fukuoka Asian Art Museum, Fukuoka, Japan
1999   Kunsthalle Bern, Projektraum (Katalog)
1998   Artsonje Center, Seoul (Katalog)
1997   Projects, The Museum of Modern Art, New York
1988   IL Gallery, Seoul

GRUPPENAUSSTELLUNGEN (AUSWAHL)
2000   Heinrich Anton Müller / Lee Bul, Bawag Foundation, Wien (Katalog)
2000   Sons et lumière, Centre Georges Pompidou, Paris (Katalog)

2000	Echigo-Tsumari Art Triennial, Echigo-Tsumari, Japan (Katalog)
2000	Let's Entertain, Walker Art Center, Minneapolis / Portland Art Museum, Portland / Saint Louis Art Museum, Saint Louis / Rufino Tamayo Museum, Mexico City (Katalog)
1999	Zeitwenden, Kunstmuseum Bonn / Museum moderner Kunst Stiftung Ludwig, Wien (Katalog)
1999	The Anagrammatical Body, Kunsthaus Muerz, Mürzschlagg / Zentrum für Kunst und Medientechnologie, Karlsruhe (Katalog)
1999	La casa, il corpo, il cuore, Museum moderner Kunst Stiftung Ludwig, Wien / National Gallery, Praha (Katalog)
1999	dAPERTutto, La Biennale di Venezia (Katalog)
1999	The Korean Pavillion, La Biennale di Venezia (Katalog)
1999	Hot Air, Granship Center, Shizuoka, Japan (Katalog)
1999	Cities on the Move 4, Louisiana Museum of Modern Art, Humlebæk, Denmark / Hayward Gallery, London / Kiasma Museum of Contemporary Art, Helsinki (Katalog)
1998–1999	Slowness of Speed, National Gallery of Victoria, Melbourne / Art Gallery of New South Wales, Sydney (Katalog)
1998	Sarajevo 2000, Museum moderner Kunst Stiftung Ludwig, Wien
1998	The Natural World, Vancouver Art Gallery, Vancouver
1998	Hugo Boss Prize, Guggenheim Museum Soho, New York (Katalog)
1997–1998	Cities on the Move, Wiener Sezession, Wien / Cities on the Move, capc Musée d'art contemporain, Bordeaux / P.S.1 Museum, New York (Katalog)
1997	Fast Forward, The Power Plant Contemporary Art Center, Toronto
1997	L'autre, 4th Biennale de Lyon (Katalog)
1996	Join Me!, Spiral / Wacoal Art Center, Tokyo
1995	Information and Reality, Fruitmarket Gallery, Edinburgh (Katalog)
1995	6. Triennale Kleinplastik 1995: Europa-Ostasien, Stuttgart (Katalog)
1995	Kwangju Biennale, Kwangju (Katalog)
1995	Ssack, Seoul Sonje Museum, Seoul (Katalog)
1994	Technology, Information, and Environment, Recycling Art Pavilion, Expo Science Park, Taejon (Katalog)
1994	The Vision of the Next Generation, Seoul Arts Center, Seoul (Katalog)
1994	This Kind of Art – Dish Washing, Kumho Museum Seoul (Katalog)
1994	Woman: The Difference and the Power, Hankuk Museum, Seoul und Yongin (Katalog)
1993	Asia-Pacific Triennial of Contemporary Art, Queensland Art Gallery, Brisbane (Katalog)

PERFORMANCES

1993	Conversation, Proto Theater, Tokyo
1993	Impromptu Amusement, Kunitachi Art Hall, Tokyo
1992	Diet: Diagramming III, Sagak Gallery, Seoul
1992	Years of Ears: Diagramming II, Live House Nanjang, Seoul
1990	Sorry for suffering – You think I'm a puppy on a picnic?, 12-day performance, Kimpo Airport / Narita Airport / downtown Tokyo / Dokiwaza Theater, Tokyo
1990	The Song of the Fish, Dong Soong Art Center, Seoul
1989	Cravings, National Museum of Contemporary Art, Seoul
1989	Abortion, Dong Soong Art Center, Seoul

AUSGESTELLTE WERKE

— *Cyborg W 1*, 1998, Silicone, polyurethane filling, paint pigment, 185 x 56 x 58 cm

— *Cyborg W 2*, 1998, Silicone, polyurethane filling, paint pigment, 185 x 74 x 58 cm
— *Cyborg W 3*, 1998, Silicone, polyurethane filling, paint pigment, 185 x 81 x 58 cm
— *Cyborg W 4*, 1998, Silicone, polyurethane filling, paint pigment, 188 x 60 x 50 cm

Martina Klein
1. Mai – 20. Juni 1999

Martina Klein

Geboren 1962, lebt und arbeitet in Düsseldorf
1982–1988   Studium der Malerei in Kassel

EINZELAUSSTELLUNGEN (AUSWAHL)

1999	Galerie Konrad Fischer, Düsseldorf
1999	Kunsthalle Bern, Projektraum (Katalog)
1998	Arnaud Lefebvre Galerie, Paris
1997	Slewe Galerie, Amsterdam (Katalog)
1997	Galerie Tschudi, Glarus
1996	Arnaud Lefebvre Galerie, Paris
1995	Galerie Konrad Fischer, Düsseldorf
1995	Slewe Galerie, Amsterdam
1994	Arnaud Lefebvre Galerie, Paris
1993	Arnaud Lefebvre Galerie, Paris
1993	Carine Campo Galerie, Antwerpen
1992	Galerie Konrad Fischer, Düsseldorf

GRUPPENAUSSTELLUNGEN (AUSWAHL)

1999	Slewe Galerie, Amsterdam
1999	Galerie Tschudi, Glarus
1998	Monomania, Rocket Galerie, London
1997	Voix, Arnaud Lefebvre Galerie, Paris
1997	Abstraits, 4 artistes au Quartier, Centre d'Art Quimper (Katalog)
1996	Star Projects Nr. 4, Vrieshuis Amerika, Amsterdam
1996	C. Delank Galerie, Bremen
1996	Arnaud Lefebvre Galerie, Paris
1996	Slewe Galerie, Amsterdam
1995	Notes on Print with and after Robert Morris, Cabinet des Estampes, Genève (Katalog)
1995	Dem Herkules zu Füßen, Museum Fridericianum, Kassel
1995	Das Abenteuer Malerei, Kunstverein Düsseldorf / Stuttgart
1995	Une Constellation, Arnaud Lefebvre Galerie, Paris
1993	Devoirs de Vacances, Arnaud Lefebvre Galerie, Paris

AUSGESTELLTE WERKE

— *Ohne Titel* (ocker dunkel / hell blau), 1998, Öl auf Leinwand, 170 x 170 x 170 cm
— *Ohne Titel* (rot grau / weiß grau), 1999, Öl auf Karton, 30 x 30 x 30 cm
— *Ohne Titel* (gelb hell / gelb hell), 1999, Öl auf Leinwand, 195 x 195 x 195 cm
— *Ohne Titel* (braun / beige), 1999, Öl auf Leinwand, 195 x 195 x 195 cm
— *Ohne Titel* (braun dunkel / rot), 1999, Öl auf Karton, 30 x 30 x 30 cm
— *Ohne Titel* (grün hell / rot dunkel), 1998, Öl auf Leinwand, 80 x 80 x 80 cm
— *Ohne Titel* (ocker hell / orange), 1998, Öl auf Leinwand, 170 x 170 x 170 cm
— *Ohne Titel* (weiß / schwarz), 1998, Öl auf Leinwand, 80 x 80 x 80 cm

@home
Natural Pollination
2. Juli – 22. August 1999

@home

1998	Soirée 01–04 @home (Pop-Events) >@home >Liege und Sitzinstallationen mit Ambientmusik und Lichtspielen >Projektleitung: @home-Crew >Coproduktion mit Theater Klara!
1998	@home Pilo-Intervention (Aktion) >Galerie Littmann ART 98 >Installation mit Pilo-Elementen und Honigmilchausschank >Projektleitung: @home-Crew
1998	m.l.stromprod zeigt: exmp.on.:knd.fir. (Video) >@home >Projektleitung: m.l.stromprod
1998	Hometrainers @ Safe (Aktion) >Art Club 98, ehem. Volksbank Basel >Konzert mit Ski Suisse >Projektleitung: Mona und Paco Manzanares
1998	Flugschule by transphere® (Labor) >Sudhaus Kulturzentrum Warteck P.P., Basel >Environment mit ferngesteuerten Luftschiffen, Pilo-Elementen, Licht und Diaprojektionen auf schwebende Kugeln, interaktivem Soundsystem, Live-Elektronik und Hare-Krishna-Essen >Projektleitung: Mona, Stefan Feger, Hansjörg Walter >Mit Christina Hagmann, Simone Fuchs, Ulrich Muchenberger, Daniel Meier, Lukas Gloor und Erdwerk
1998	Wax @home (Workshop und Aktion) >@home >Kerzenziehinstallation, Sound und Teeausschank >Projektleitung: Christina Hagman, Simone Fuchs, Gini Messerli
1999	Mondmilchfahrt (Poetisches Reiseevent, Roadmovie No. 01) >Zürich–Basel–La Chaux-de-Fonds–Lac des Taillères >Carfahrt und Outdoorhappening im Schnee >Projektleitung: Mona >Coproduktion mit leparadis.chx
1999	Natural Pollination (Installation) >Projektraum Kunsthalle Bern >Wandzeichnungen, Kristalle, Licht, Luftbefeuchter, Duftessenz, Sitzkissen, Tonerzeuger, Faxe >Projektleitung: Simone Fuchs, Ira Schulthess, Mona, Christina Hagmann, Klara Borbely, Maat
1999	Die diplomatische Vertretung (Environment) >Ausstellung Get Together, Kunsthalle Wien >Pneumatische Raumarchitektur, Tierfelle, Lautsprecher, Handy, Messingobjekt und Katalogbeitrag >Projektleitung: Mona, Christina Hagmann, Stefan Feger
1999	Schallwellen (Audioforum) >Viper 99, Casino Bar Luzern >Projektleitung: Dominik Ziliotis und Freunde
1999	Integral No. 04 Knut und Sylvie @home (Plattentaufe) >@home >Projektleitung: @home-Crew >Mit Blume und Beat Brogle
1999	Integral No. 05 Ski Suisse @home (Plattentaufe) >@home >Projektleitung: Dominik Ziliotis und Freunde

AUSGESTELLTE WERKE

— *Natural Pollination,* Installation (Wandzeichnungen, Kristalle, Licht, Luftbefeuchter, Honigessenz, Sitzkissen, Tonerzeuger, Faxe)

Heinz Brand
18. September – 3. Oktober 1999
Eine Ausstellung des Projektraums in Ornans (F)

Heinz Brand

Geboren 1944, lebt und arbeitet in Zollikofen und Bern

1991–1992	P.S.1, New York
1990	Schweizer Institut, Roma
1987	Atelierstipendium an der Cité Internationale des Arts, Paris
1984	Studienaufenthalt in Kyoto
1976–1978	Auslandstipendium an der Jan van Eyck Akademie, Maastricht
1974–1975	Auslandstipendium an der Staatlichen Hochschule für Bildende Kunst in Lodz
1969–1971	Rio de Janeiro
1968	Fotoklasse Zürich und F+F Zürich
1966	Studienaufenthalt in Kyoto
1960–1965	Kunstgewerbeschule Bern

EINZELAUSSTELLUNGEN (AUSWAHL)

1999	Kurzbiografie, Galerie Stampa, Basel
1999	Kunsthalle Bern, Projektraum (Katalog)
1999	Gold light district, Neuer Kunstverein, Gießen
1998	Gleichzeitig, Schloß Reichenbach, Zollikofen
1997	Weihnachts-Installation, Bern
1997	Fernrohr, Bern
1997	Plötzlich ist alles ganz anders: 3 Installationen zum Andenken an Irma Brand (1912–1995), Schloß Reichenbach, Zollikofen
1996	Seitenverkehrte Abwesenheit, Galerie Stampa, Basel
1994	Koan, Toni Gerber, Bern
1994	Blind cave fish in the dark, Galerie Francesca Pia, Bern
1992	Intermission between nothing, P.S.1, Room 201, New York
1991	Hier und weg: Räume, Städtische Galerie am Markt, Schwäbisch Hall (Katalog)

GRUPPENAUSSTELLUNGEN (AUSWAHL)

1998	Der Niesen: Ein Berg im Spiegel der bildenden Kunst, Kunstmuseum Thun (mit Fritz Brand; Katalog)
1996	Ohne Titel: Eine Sammlung zeitgenössischer Kunst: Stiftung Kunst Heute, Aargauer Kunsthaus, Aarau (Katalog)
1996	Die Sammlung Toni Gerber im Kunstmuseum Bern, 2. Teil, Kunstmuseum Bern (Katalog)
1996	Im Kunstlicht: Photographie im 20. Jahrhundert aus den Sammlungen im Kunsthaus Zürich, Kunsthaus Zürich (Katalog)
1996	Wege des Imaginären in der Wiedergabe des Realen: Photographie & die Folgen, Untere Fabrik, Sissach
1995	Schweizerische Plastik-Ausstellung, Môtiers (Katalog)
1995	Die Sammlung Schweizer Photographie der Gotthard Bank, Galleria Gottardo, Lugano / Galleria Matasci, Tenero / Centro d'arte contemporanea, Bellinzona (Katalog)
1994	Berner Biennale 1994, Kunsthaus Langenthal (Katalog)
1992	Encounters with diversity, P.S.1, Museum, New York (Katalog)
1992	Tierra de Nadie, Granada (Katalog)

1991   Ex Aequo: 24 Schweizer Künstler in St-Imier, St-Imier (Katalog)
1990   Blau – Farbe der Ferne, Heidelberger Kunstverein, Heidelberg (Katalog)

AUSGESTELLTE WERKE
— *Installation o.T.*, 1999, Außenraum (250 x 250 x 400 cm), Tisch (Durchmesser 100 cm),
Photographie (58 x 71 cm), ein offener Raum mit drei imaginären Wänden, gereinigt
wurde, was gereinigt werden konnte
— *Ein Weltbild in ein anderes hineingestellt*, 1999, Hotelzimmer (200 x 250 x 300 cm),
Marmorplatte (92 x 40 x 1 cm), *Whites* (Rückseite einer Photographie, 59,5 x 71,5 cm),
Whites (Rückseite einer Photographie, 84 x 59,5 cm)
— *Unsichtbares Selbstbildnis*, 1986/99, Sockel (Durchmesser 70 cm), Scharfeinstellung

Eran Schaerf
Scenario Data
16. Oktober – 5. Dezember 1999

Eran Schaerf

Geboren 1962, lebt in Brüssel und Berlin
1985–1987   Hochschule der Künste, Berlin
1978–1982   Technikum, Giv'ataim

EINZELAUSSTELLUNGEN (AUSWAHL)
1999   Scenario Data, Zwinger Berlin
1999   Scenario Data, Kunsthalle Bern, Projektraum (Katalog)
1998   Continuity of Breaks, Titanik, Turku (Katalog)
1998   Scenario Data, Herzliya Museum of Art (Katalog)
1998   Scenario Data, Ecole des Beaux Arts de Nantes
1997   Recasting, Museum van Hedendaagse Kunst Antwerpen / Kunstverein
            München / FRAC Champagne-Ardenne, Reims (Katalog)
1996   Re-enactment, Bahnwärterhaus, Esslingen a. N. (Katalog)
1995   Touch-Wood, Galerie Barbara Gross, München
1995   Zaun-Town, Portikus, Frankfurt am Main (Katalog)
1994   We is o.k., De Vleeshal, Middelburg
1994   Home Science, Zwinger, Berlin
1994   Wall-like Law, Stedelijk Museum, Bureau Amsterdam,
            mit Jan van Grunsven
1993   Ciel et ciel, Vereniging voor het Museum van Hedendaagse Kunst, Gent
1992   Ciel, Zwinger, Berlin
1991   Schneider u. Sohn, längen, Kürzen, Rosen, Zwinger, Berlin
1990   Einladungskarte, Bestellkarte, Bachstelze, Künstlerhaus Bethanien, Berlin
1988   (It's) I prefer chocolate, Galerie Anselm Dreher, Berlin

FILME (MIT EVA MEYER)
1999   *Europa von weitem*, Video, 73 Min. 46 Sek.
1998   *Documentary Credit*, Video, 71 Min.
1997   *Wie gewohnt: Ein Versatzstück*, Video, 26 Min. 40 Sek.

HÖRSPIELE
1999   *Europa von weitem*, Bayerischer Rundfunk, München, mit Eva Meyer
1997   *Wie gesagt, Theater- oder Taxistück*, Bayerischer Rundfunk, München (CD)

AUSGESTELLTE WERKE

— *Scenario Data # 18 – # 31* (Eintrittskarten), 1999, Text: Eran Schaerf, Design: salary-man und Eran Schaerf, Offset auf Papier, je 19 x 6 cm, Reihenfolge variabel

— Scenario Data # 32, 1999, Diapositive, Projektoren, Drehbühnen, Holzplatten, Böcke, Dimension variabel – Mit Dank an: Micha Bar-Am / Magnum: *Last Day of the Yom Kippur War, West Side of the Suez Canal,* 1974 / Stanley Kubrick: *2001 – A Space Odyssey,* 1968 / Israel Museum Jerusalem / David Harris: *Das Rothschild Zimmer,* Französischer Salon des 18. Jahrhunderts, 1969 / Israel Museum Jerusalem: *Baron Edmond de Rothschild weiht das Zimmer im Israel Museum ein,* 1969 / Israel Museum Jerusalem: *Das Rothschild Zimmer,* Französischer Salon des 18. Jahrhunderts, Sessel, 1969 / *Europa Militaria,* Special Number 12: *Napoleon's Imperial Guard, Waterloo, Re-enactment,* undatiert / *Europa Militaria,* Special Number 9: *German Napoleonic Armies, Jena, Wagram, Leipzig and Waterloo, Re-enactment,* undatiert / Anonym: *Sechs-Tage-Krieg, Israelische Soldaten vor einem Porträt von Nasser,* 1967 / Archiv der Monatsschrift der Israelischen Armee: *Yakoba als Araber verkleidet auf einer Straße in Damaskus,* 1948 / *Yakoba in der Uniform der arabischen paramilitärischen Jugend-Organisation,* undatiert / *Oberkommandant der Israelischen Armee bei einer Besprechung,* undatiert / *Oberkommandant der Israelischen Armee als Samson-Araber,* 1991 / *Soldaten der Samson-Truppe vor einer Aktion,* undatiert / *Soldaten der Samson-Truppe,* ohne Titel, undatiert / *Soldat der Truppe Kirsche* (gemäß *New York Times*) *als Araber verkleidet,* undatiert / *Angehörige der Samson-Truppe als Araberin und Araber verkleidet,* undatiert

Anya Gallaccio
29. Januar – 12. März 2000

Anya Gallaccio

Geboren 1963, lebt und arbeitet in London

1999	Hamlyn award, Paul Hamlyn Foundation
1998	Sargent Fellowship, British School at Rome
1997	Art Pace International Artist's Programme, San Antonio, Texas
1985–1988	Goldsmith's College, University of London
1984–1985	Kingston Polytechnic

EINZELAUSSTELLUNGEN (AUSWAHL)

2000	Kunsthalle Bern, Projektraum (Katalog)
1999	Glaschu, Tramway at Lanarkshire House, Glasgow (Katalog)
1998	Bloom Gallery, Amsterdam (Katalog)
1998	Two Sisters, Minerva Basin, Hull (for Locus + and Artranspennine)
1998	Chasing Rainbows, Delfina Studios, London
1997	Blum and Poe, Los Angeles
1997	Serpentine Gallery Lawn, London
1997	Art Pace, San Antonio, Texas (Katalog)
1996	A Multiple, Riding House Editions, London
1996	Galerie Rodolphe Janssen, Bruxelles
1996	Galerie im Künstlerhaus, Bremen (Katalog)
1996	Ars Futura Galerie, Zürich
1996	Intensities and Surfaces, Wapping Pumping Station, London (for the Women's Playhouse Trust)
1995	Towards the Rainbow, Angel Row Gallery, Nottingham
1995	Stephen Friedman, London
1994	stroke, Blum and Poe, Los Angeles
1994	Couverture, Filiale, Basel (Katalog)
1994	stroke, Karsten Schubert, London

1994	La dolce vita, Stephania Miscetti, Roma
1993	Galerie Krinzinger, Wien (Katalog)
1993	Ars Futura Galerie, Zürich
1993	Kim Light Gallery, Los Angeles
1992	red on green, ICA, London
1991	Karsten Schubert, London

GRUPPENAUSSTELLUNGEN (AUSWAHL)

1999	Prime, Dundee Contemporary Arts
1999	Releasing the senses, Tokyo Opera City, Japan (Katalog)
1999	Viereck und Kosmos, Amden
1998	Real Life: New British Art, British Council touring exhibition, Japan (Katalog)
1997	Pictura Britannica, MOCA, Sydney
1996	The Pleasure of Aesthetic Life, The Showroom, London
1996	Private View: Contemporary British and German Artists, A New Collection for John and Josephine Bowes, The Bowes Museum, Barnard Castle, Durham
1995	The British Art Show 4, South Bank touring exhibition, Manchester, Edinburgh, Cardiff (Katalog)
1995	Brilliant, Walker Art Center, Minneapolis / CAM Houston, Texas
1995	Where you were even now, Kunsthalle Winterthur (Katalog)
1995	Chocolate!, The Swiss Institute, New York (Katalog)
1994	Art Unlimited: Multiples from the 1960s and 1990s, South Bank Centre touring exhibition (Katalog)
1994	InSITE, 94, MOCA, San Diego / Agua Caliente, Tijuana, Mexico
1994	Domestic Violence, Gio Marconi, Milano
1993	Le Jardin de la Vierge, Musée Instrumental, Bruxelles
1993	Le Principe de Réalité, Villa Arson, Nice
1992	With Attitude, Galerie Rodolphe Janssen, Bruxelles
1992	Sweet Home, Oriel Mostyn, Gwynedd
1992	Barbara Gladstone Gallery and Stein Gladstone Gallery, New York
1992	Life Size, Museo d'Arte Contemporanea, Prato
1991	Confrontaciones, Palacio de Velazquez, Madrid
1991	Broken English, Serpentine Gallery, London (Katalog)
1988	Freeze, Surrey Docks, London

AUSGESTELLTE WERKE

— *falling from grace*, 2000, Äpfel / Schnur, 4,5 x 12,5 m

Renate Buser   Katharina Grosse
Marie Sester   Susanne Fankhauser
Elizabeth Wright   Sarah Rossiter

146

Lee Bul
Martina Klein
@home

Heinz Brand
Eran Schaerf
Anya Gallaccio

Zum Autor / The Author

Roman Kurzmeyer, geb. 1961, Dr. phil., Arbeitsschwerpunkte bilden Untersuchungen zum Einfluß von Lebens- und Produktionsbedingungen auf die künstlerische Praxis im 20. Jahrhundert, die Erarbeitung von Ausstellungen und Publikationen sowie Texte zur zeitgenössischen Kunst. Dozent an der Hochschule für Gestaltung und Kunst Basel. 1998–2000 Leiter des Projektraums der Kunsthalle Bern.

Roman Kurzmeyer, Ph. D. (1961–). His research has focused on the influence of living conditions and production circumstances on the practice of art in the 20th century. He has curated a number of exhibitions and authored various publications and texts on contemporary art. Roman Kurzmeyer is a lecturer at the Hochschule für Gestaltung und Kunst Basel. He served as director of the "Projektraum" at the Kunsthalle Bern from 1998 to 2000.

Publikationen (Auswahl)
Selected Publications

— Bettina Hunger / Michael Kohlenbach / Roman Kurzmeyer / Ralph Schröder / Martin Stingelin / Hubert Thüring (Hgg.), *Adolf Wölfli: Porträt eines produktiven Unfalls – Dokumente und Recherchen*, Basel / Frankfurt am Main 1993.

— Roman Kurzmeyer (Hg.), *Fritz Pauli (1891–1968): Maler und Radierer*, Basel / Frankfurt am Main 1994.

— Roman Kurzmeyer (Hg.), *Heinrich Anton Müller (1869–1930): Katalog der Maschinen, Zeichnungen und Schriften*, Basel / Frankfurt am Main 1994.

— Roman Kurzmeyer, *Annelies Štrba: Ware Iri Ware Ni Iru*, Luzern 1994.

— Roman Kurzmeyer, *Zwei Zimmer für Julie Bondeli: Barry Ratoff, Eva Schlegel, Hannah Villiger, Elizabeth Wright*, Bern 1996.

— Roman Kurzmeyer, *Auf ein Bild hin: Jakob Wäch (1893–1918)*, Basel / Frankfurt am Main 1997.

— Josef Helfenstein / Roman Kurzmeyer (Hgg.), *Bill Traylor (1854–1949): Deep Blues*, Köln 1998.

— Roman Kurzmeyer, *Viereck und Kosmos: Künstler, Lebensreformer, Okkultisten, Spiritisten in Amden 1901–1912: Max Nopper, Josua Klein, Fidus, Otto Meyer-Amden*, Wien / New York 1999.

Ines Anselmi / E. Valdés Figueroa (Hgg.)

Neue Kunst aus Kuba
Art actuel de Cuba
Arte cubano contemporáneo

La dirección de la mirada

Deutsch / Français / Español
144 Seiten, 82 Abb., geb., Fadenheftung
CHF 36 / DEM 39 / ATS 273
ISBN 3-211-83301-3

Die »Neue Kubanische Kunst« der 90er Jahre ist Markenzeichen, tolerierter Skandal und bilderreicher Dissens in einem. In »elitärer« Vieldeutigkeit formuliert, entzieht sich die bildnerische Produktion einer neuen Generation spielerisch der lokalen Zensur und sichert sich gleichzeitig die Aufmerksamkeit der Kuratoren aus Los Angeles und Madrid.

Texte von Jorge Ángel Pérez, Ines Anselmi, Abelardo Mena, Gerardo Mosquera, Eugenio Valdés Figueroa.

»New Art from Cuba« of the nineties is three things in one: trade-mark, tolerated scandal, and art-inspiring dissent. Deliberately claiming an »elitist« position of ambiguity this young generation of artists manages to outwit local censorship as well as to attract the attention of curators in Los Angeles or Madrid.

»[...] die Möglichkeit zu entdecken, daß zeitgenössisches Kunstschaffen von Fidel Castros Insel weder mit agit-properen Polit-Pamphleten noch mit den Stereotypen eines folkloristisch angehauchten ›lateinamerikanischen Phantastischen‹ etwas zu tun hat. Vielmehr präsentiert sich ein Panoptikum verblüffend moderner Exponate [...] ein sehr schön gestaltetes Buch.« *(Neue Zürcher Zeitung)*

Tania Bruguera / Los Carpinteros / Sandra Ceballos / Luis Gómez / Rodolfo Llópiz / Kcho / Ibrahim Miranda / Antonio Núñez / Sandra Ramos / Fernando Rodríguez y Francisco de la Cal / René Francisco Rodríguez / Lázaro Saavedra / Ezequiel Suárez / Tonel

Edition Voldemeer Zürich
Springer Wien New York

Roman Kurzmeyer

Viereck und Kosmos

Künstler, Lebensreformer, Okkultisten,
Spiritisten in Amden 1901–1912
Max Nopper, Josua Klein, Fidus,
Otto Meyer-Amden

Deutsch
264 Seiten, 80 Abb., geb., Fadenheftung
CHF 45.50 / DEM 49 / ATS 343
ISBN 3-211-83371-4

Im Jahr 1903 kaufte Josua Klein in Amden, einer am Walensee in der Schweiz gelegenen Berggemeinde, für viel Geld zahlreiche Häuser, Wiesen, Äcker und Wälder. Josua Klein, Sohn eines freireligiösen Privatgelehrten, und Max Nopper, ein aus Gewissensgründen aus dem Dienst der Württembergischen Armee ausgetretener Offizier, ließen die erworbenen Häuser reparieren und eine große Scheune und einen Zentrumsbau errichten. Josua Klein plante in Amden Tempel errichten zu lassen und engagierte für diese Aufgabe den Berliner Künstler Fidus. Unter deutschen Lebensreformern hieß es, in Amden entstehe ein Gottesreich auf Erden.

In 1903, Josua Klein spent a fortune buying numerous houses, meadows, fields, and forests in and around Amden, a small mountain village above Lake Walenstadt in central Switzerland. Klein, the son of a non-denominational scholar, together with Max Nopper, ex-officer of the Wurttemberg army and a conscientious objector, organized the restoration of the existing houses as well as the construction of new buildings, among them a number of temples to be designed by Fidus, an artist from Berlin. Among other German adherents to the lifereform-movement there was talk of a Divine Kingdom on earth being created at Amden.

»Kurzmeyer hat die Faktengeschichte [...] minuziös recherchiert und [...] Kontexte (Lebensreform, ›Tempelkunst‹) des Projekts von Nopper und Klein in dem ästethisch sehr ansprechend gestalteten Band dargestellt« *(Neue Zürcher Zeitung)*

Edition Voldemeer Zürich
Springer Wien New York

Jörg Huber / Martin Heller (Hgg.)

**Konstruktionen Sichtbarkeiten
Interventionen 8**

Deutsch
Hochschule für Gestaltung und Kunst Zürich /
Museum für Gestaltung Zürich
296 Seiten, 32 Abbildungen, Französische
Broschur, Fadenheftung
CHF 36 / DEM 39 / ATS 273
ISBN 3-211-83300-5

Konstruktionen Sichtbarkeiten: Fragen der
Herstellung, Bedeutung und Funktion von
Bildern und Visualisierungsprozessen stehen
im Mittelpunkt des vorliegenden Bandes. Beiträge aus unterschiedlichen Wissens- und
Forschungsbereichen thematisieren in diesem Zusammenhang insbesondere Fragen
der Wahrnehmung und des Erkennens, der
Realitäts- und der Existenzvergewisserung.
Sowohl geistes- als auch naturwissenschaft-
liche Perspektiven und Positionen werden
vorgestellt und verhandelt.

*Construction Visibility: How are images and
processes of visualisation produced, how do
they function, and what do they signify –
these questions are addressed in this volume
that brings together contributions from a
varied range of fields (humanities as well as
the natural sciences) focussing on the ways
we discern, perceive, and assure ourselves of
our existence and reality.*

Gottfried Boehm
Elena Esposito
Alois M. Haas
Alois Hahn
Andreas Huyssen
Karin Knorr Cetina
Joachim Krug
Nabil Mahdaoui
Ursula Panhans-Bühler
Christoph Rehmann-Sutter
Georg Christoph Tholen
Gianni Vattimo
Michael Wetzel

Edition Voldemeer Zürich
Springer Wien New York

Jörg Huber (Hg.)

**Darstellung : Korrespondenz
Interventionen 9**

Deutsch
Institut für Theorie der Gestaltung und Kunst
(ith) an der HGK Zürich
248 Seiten, 20 vierfarbige Abbildungen,
Französische Broschur, Fadenheftung
CHF 36 / DEM 39 / ATS 273
ISBN 3-211-83477-X

Darstellung : Korrespondenz. Nachdem im
Rahmen der *Interventionen* wiederholt und
in verschiedenen Kontexten die Aushebelung
der Individuen und der Kollektive beschrieben und analysiert wurde, diskutieren die
Beiträge des vorliegenden Bandes Fragen der
Verbindungen und der Verbindlichkeiten. Wie
werden Ankoppelungen und Bezugnahmen
im Gesellschaftlichen hergestellt, im Ästhetischen sichtbar gemacht, in der Theorie reflektiert?

*Representation : Communication. Interven-
tions, a series of lectures and conferences
regularly organized by the Zurich Institute of
Art and Design Theory (ith), have one of their
focuses on describing and analysing phenomena of uprootedness affecting individuals
and groups in modern society. The most recent papers published in the present volume
discuss issues of communication and commitment and investigate how, in a social
environment, connections and relations are
established, how these are visualized aes-
thetically and reflected theoretically.*

Hans Belting
Iain Chambers
Ulla Haselstein
Axel Honneth
Huang Qi
Mario G. Losano
Herta Nagl-Docekal
Barbara Maria Stafford
Christoph Wulf
Beat Wyss

Edition Voldemeer Zürich
Springer Wien New York